ISBN 978-0-266-15545-4
PIBN 10020325

MODEL MAKING.

INCLUDING

WORKSHOP PRACTICE, DESIGN AND CONSTRUCTION OF MODELS

A PRACTICAL TREATISE FOR THE AMATEUR AND PRO-
FESSIONAL MECHANIC—GIVING INSTRUCTIONS ON THE
VARIOUS PROCESSES AND OPERATIONS INVOLVED
IN MODEL MAKING AND THE ACTUAL CONSTRUC-
TION OF NUMEROUS MODELS, INCLUDING
STEAM ENGINES, SPEED BOATS, GUNS,
LOCOMOTIVES, CRANES, ETC.

LATHE WORK, PATTERN WORK, ELECTROPLATING,
SOFT AND HARD SOLDERING, GRINDING,
DRILLING, ETC., ARE ALSO INCLUDED

EDITED BY

RAYMOND FRANCIS YATES

EDITOR OF "EVERYDAY ENGINEERING MAGAZINE"

FULLY ILLUSTRATED WITH 300 ENGRAVINGS

REPRODUCTIONS OF ACTUAL WORKING MODELS

NEW YORK
THE NORMAN W. HENLEY PUBLISHING COMPANY
2 WEST 45th STREET
1919

PRINTING, PRESSWORK AND BINDING BY THE
PUBLISHERS PRINTING CO., NEW YORK CITY

PREFACE

Model making is far from a senseless hobby—just the opposite; it is practical, educating and carries with it the prestige and dignity of a specialized science. Its scope is unlimited and its ramifications are unnumbered. The giant four-cylinder compound locomotive is reproduced in miniature complete in every detail from Walschaert valve to throttle; a torpedo boat destroyer is modeled and provided with workable steam engines, a model speed boat is constructed and coaxed into going 30 miles per hour; a six-cylinder engine is built with a six-throw crankshaft turned out of a solid piece of steel. This cannot exactly be called model making. The expression is inadequate and does not carry with it the full meaning of the work. It is really model engineering—engineering in miniature. The construction of a model locomotive involves no small amount of work and knowledge. Its constructor must know something of steam engineering, he must be able to read the most advanced blueprints to enable him to produce his model to scale from a drawing of its prototype. Aside from this, he must be a mechanic of no mean ability. He must possess infinite patience and resourcefulness. Of course, not every model maker can build a locomotive. More simple mechanisms are usually chosen to start with. This is where part of the real value of model making presents itself, and its educating value becomes manifest. The man who makes a miniature locomotive, a torpedo boat destroyer or airship, has increased his own knowledge to a great extent; the experience has made him a better mechanic. In many cases, the fundamental

principles of operation must be mastered before the
model is made. As an example: A young man decides
to make a workable model of a gasoline engine. First,
unless already acquainted with its principles of operation, he must study them until he becomes sufficiently
acquainted with them to proceed intelligently with the
design and construction of his machine. The engine
must be carefully laid out and drawn accurately to scale;
its bore, stroke, power and cycle must all be decided
upon. After the design is completed upon paper, the
patterns for its castings must be turned out and then
the machining starts. Precision and accuracy is essential to a well-working engine and the lathe must be
manipulated with skilful fingers. The engine is finished
and assembled. What has its builder accomplished? He
is perfectly satisfied to stand and watch it run on the
workshop bench. That is all he made it for, but aside
from this, the love of his hobby has taught him much
of practical value, as can readily be understood. The
thousands of model makers in England have been of
great value to their country through the wonderful
knowledge they obtained by "tinkering" with models.

After a man spends many hours — yes, even days,
weeks and months—on the model of a certain machine,
upon completion the thing represents something to him
very remote from money. It is not made for money, and
therefore its value is not estimated in money. It is difficult to explain just how a man regards his model. His
eyes never tire of it—he actually loves it. The writer
has in mind a man who cried like a child when a model
upon which he had worked faithfully for a period of
many months was damaged beyond repair in transportation. The man was no exception. The insurance he
received from the express company was nothing to him
in comparison to his work. He was merely an ordinary
modeler possessed of some peculiar God-given instinct

that made him love a miniature creation of his own hands.

The reader will notice an unmistakable defensive tone in the above paragraphs. True, it is defensive and intended largely to vindicate the fact that a modeler is not necessarily an immature youth or a man with childish notions and that model making is an engineering science rather than a senseless hobby.

In England, model making or model engineering carries with it a different meaning than it does at the present time in America. There has been a tendency in this country to associate it with toy making, and no comparison could be more vague or humiliating to the ardent followers of the work than this. It is utterly unjustified and the impression remains merely because the public in this country has not been educated to the true meaning and significance of the science—a category in which it rightfully has a place. It is hoped that this book, which is prepared to give impetus to model engineering in this country, will also serve the double purpose of creating a correct impression of the work its pages are devoted to.

The best articles on model engineering that have appeared in *Everyday Engineering Magazine* during the past two years have been chosen for this book. It was the writer's task to re-edit these articles and arrange them for publication, together with 29,000 words of his own writing relative to model engineering, part of which have appeared in past issues of the magazine and part of which is being published for the first time. Due acknowledgment and thanks are given to the various other contributors, as follows: J. F. Springer, H. H. Parker, George Bender, Thomas S. Curtis, Arthur J. Weed, A. Koster, L. F. Carter, Ian McKinzie, Ralph R. Weddell, Raymond Tetens, James E. Carrington, J. Fawcett Rapp, Omer Cote, S. H. Kershaw, and E. B. Nichols. The writer wishes to acknowledge his especial indebtedness to

Mr. George Bender for his valued assistance and suggestions in connection with the actual preparation of the volume. Mr. Bender is one of the most enthusiastic and able model engineers in the country and the truth of this statement is well substantiated by the descriptions of some of his models which appear in this volume, especially the model steam locomotive in Chapter XXVIII.

If this book helps to create interest in model engineering and also helps to correct the impression of a misinformed public relative to the science it treats, the highest wishes of those responsible for its inception will be realized.

RAYMOND FRANCIS YATES.

January, 1919.

CONTENTS

CHAPTER I

THE MODEL ENGINEERS' WORKSHOP

CHAPTER II

LATHES AND LATHE WORK

CHAPTER III

DRILLS AND DRILLING

CHAPTER IV

SOFT AND HARD SOLDERING

CHAPTER V

HARDENING AND TEMPERING STEEL

CHAPTER VI

THE USE OF ABRASIVES

CHAPTER VII

PATTERN MAKING

CHAPTER VIII

ELECTRO-PLATING

CHAPTER IX

A MODEL ·SLIDE CRANK STEAM ENGINE

CHAPTER X

A MODEL TWIN-CYLINDER ENGINE

CHAPTER XI

A SINGLE-CYLINDER ENGINE

CHAPTER XII

A MODEL TWIN-CYLINDER MARINE ENGINE

CHAPTER XIII

FLASH STEAM PLANTS

CHAPTER XIV

A FLASH STEAM PLANT FOR LARGE MODEL AIRPLANES

CHAPTER XV

A FLASH STEAM PLANT FOR SMALL MODEL AIRPLANES

CHAPTER XVI

A MODEL STEAM TURBINE

CHAPTER XVII

DESIGN AND CONSTRUCTION OF MODEL BOILERS

CHAPTER XVIII

MODEL BOILER FITTINGS

CHAPTER XIX

A RECORD-BREAKING MODEL HYDROPLANE

CHAPTER XX

A MODEL LAKE FREIGHTER

CHAPTER XXI

A SHARPIE-TYPE MODEL BOAT

CHAPTER XXII

A MODEL SUBMARINE CHASER

Contents

CHAPTER XXX

A MODEL CATERPILLAR TANK

CHAPTER XXXI

A MODEL SIEGE GUN

MODEL MAKING

CHAPTER I

THE MODEL ENGINEERS' WORKSHOP

Notes on the general arrangement of the model engineers' workshop and tool equipment—The locations of the shop, its lighting, heating, ventilation and power-driven machines.

Before considering the tool equipment and general arrangement of the model engineers' workshop, a few words will be said in regard to the location of the shop, which is of some importance. If the workshop is in the house it is generally necessary to either put it in the cellar or in the attic; both locations have their advantages and their disadvantages. The attic is extremely warm in the summer and unbearably cold in the winter. The attic does possess the advantage, however, of being very dry and, in the average case, quite light. Cellars are often quite damp and this is bad for the tool equipment, as good steel rusts very easily unless it is properly protected. The cellar, however, is delightfully cool in the summer and, with the furnace, it is comfortable in the winter. Of course, the matter of light must be considered if the shop is to be placed in the cellar, as artificial illumination must be employed and this is by no means desirable. The lighting problem can be partly overcome by giving the walls a good coating of whitewash. This reflects light and it is not necessary to employ such powerful lamps to illuminate a given space. The whitewash also makes the cellar more clean and wholesome. A wooden floor placed over the concrete is also a decided improvement and offers some protection to the worker. If a good, substantial cabinet is made to keep the tools in, the trouble from dampness will be almost entirely obviated—providing the tools are not left lying around the bench. The lathe, drill press and other small machine equipment should be kept well covered when not in use and, if left for a considerable time, the parts that are not enameled should be smeared with vaseline, which can be easily

Fig. 1—A corner in a model engineer's workshop

removed and which is very effective in preventing rust and protecting the steel from moisture.

Owing to the low ceiling in most cellars, it is almost impossible to drive the lathe and drill press or other machinery from a line shaft and the independent motor drive must be introduced. This is to be desired and recommended. Although expensive, it is more convenient and satisfactory. If second-hand motors are purchased the expense need not be prohibitive. A well-arranged shop in the cellar is shown in the illustrations, and from these the reader will get a good idea of just what can be done by using a little care and forethought in making the plans. It will be seen that everything is systematically arranged and that every tool has a place.

Fig. 2—A power-driven jig-saw in a workshop

If there is a window in the workshop, the lathe should be set up before it, as this offers the very best light possible to work by. If such an arrangement is not possible, a shaded light will have to be dropped on a cord over the lathe. In this case, it is well to arrange a small board in back of the machine and place all the gears, wrenches, attachments, etc., upon it where they will be convenient to reach. A small shelf can be placed at the bottom of the board to hold tools, reamers, drills and measuring instruments while the workman is using the machine.

If the lathe is motor-driven, the controlling switch should be placed so that the operator can reach it with little trouble, as it is oftentimes necessary to stop a lathe quickly, especially in screw-cutting. Further information on setting up a small lathe is given in Chapter II.

The work bench is an important part of the equipment of a small shop and should be given careful attention. A large bench should be made when possible, providing there is sufficient room, as a small bench soon becomes littered with tools when a job is being done, and this necessitates searching for certain tools more often than would otherwise be necessary if a larger bench were built. Two-inch planking should be used for the top of the bench. Twenty-four inches is a good width for the bench, although 30 inches does not make it too wide. The supporting posts should be made of 2 x 2-inch stock and these members should be placed not more than 3 feet apart if a good, substantial bench is desired. A long tool rack can be placed behind the bench and this can be formed by a long plank nailed or bolted to the wall. Various tools can be arranged upon this. In many cases they can be held in place by means of nails and such things as files can be held in a special rack cut from a piece of wood. Both a heavy and a light vise should be attached to the bench at opposite ends, and a light should be placed quite close to each one, as very careful and accurate work is often done in the vise and sufficient light is quite necessary. The work bench can be elaborated upon by adding several drawers to it, as there is a multitude of things that may be kept in them. A splendid addition to the bench can be made by building a shelf over it where stock of various nature can be stored out of the way. Lumber, brass rods, steel rods, metal sheeting, etc., can be placed on this shelf where it will be out of the way and yet always conveniently available.

In one corner of the shop, a small shelf should be made

upon which all soldering, brazing, tempering and heating operations in general should be done. This shelf should be well covered with asbestos board or heavy asbestos sheeting to render it absolutely fireproof. If this shelf is placed against a wooden wall, it will also be necessary to protect the wall in the same manner. In the cellar, where there is a stone wall, this procedure

Fig. 3—A tool cabinet in a cellar workshop

will not be necessary, although a small hood made from heavy asbestos sheeting should be placed over the shelf to prevent sparks from reaching the ceiling. A small bellows can be placed under the shelf, where it can be operated with the foot if the worker desires to use a slight air pressure in connection with a blow lamp or gas burner for brazing or silver soldering.

Now that the general arrangement of the model maker's workshop has been covered, the tool equipment of such a shop will be considered more in detail. Anyone can have a well-equipped shop if they have the money to invest in it, but the average model maker finds it necessary to accumulate his tools slowly and, in many cases, piece by piece, until a complete outfit is obtained. The following is a list of tools that should form the equipment of the model engineer's workshop. While many costly tools and machines could be added, the writer believes

Fig. 4—A small, power-driven drill press in a model engineer's shop

that the list is complete enough for the average fellow whose ambition in this respect is generally limited by his pocketbook:

> Small lathe (Screw-cutting if Possible)
> Drill Press
> Grinding Head (See Chapter VI)
> Hack Saw (Both Large and Small)
> Hand Drill
> Set of Drills

Gasolene Torch
Two Soldering Coppers (Large and Small)
Two Vises (Large and Small)
Assortment of Files
Micrometer
Scale
Calipers (Inside and Outside)
Six-inch Dividers
Chisels
Center Punches
Pliers
Tinner's Snips
Assortment of Taps and Dies
Tap Wrench and Die Stock
Screw Drivers
Machinists' Square
Drill Gauge
Various Hammers
Scriber

Much could be added to this list in the way of wood-working tools, but owing to the fact that most of the model engineer's work is done on metal this was not thought necessary. Two good saws, a draw-knife, a few chisels, a plane and a mitre box would compose a good wood-working equipment which would meet the needs of the average shop. Such an outfit is, of course, quite necessary if model power boat hulls are to be made. All wood turning can be done on the metal working lathe and the tools used can be ground to shape from old files, as explained in a subsequent chapter.

A very useful addition to the workshop is a small cabinet with at least twenty drawers, in which small brads, nails, tacks, screws, nuts and bolts may be kept. A sample of the screw, nut or whatever it may be that is kept in the drawer, should be fixed to the front of it so that a search for the object wanted will not be necessary. A small cabinet of this nature can be easily made by procuring a few tobacco tins of a certain size and building a small wooden rack to place them on. The tins should be provided with covers to keep dust and dirt out.

In many cases the model engineer can build his workshop in the back yard, and he can then design it so it will be well lighted, ventilated and heated. Such a shop

Fig. 5—A small engine lathe with attachments mounted on the wall where they will be convenient

should have at least 225 square feet of floor space, and more if possible. It will be quite necessary to afford ample protection from moisture and dampness when the floor is put in, and the builder is cautioned to consider this point very carefully, as a poorly constructed floor will make the place very damp and the tools will suffer greatly. Several remedies can be applied in a case like this, and probably the most practical method and the one requiring least trouble and expense is that of putting in a double floor with a layer of heavy tar-paper placed between. Many might think that raising the shop off the ground on posts would be effective in preventing trouble with moisture, and this is quite so, but, at the same time, the fact that such a building has a very cold floor in the winter time should not be lost sight of. The sides might be boarded up in the winter, but in the end this procedure

Fig. 6—A power-driven lathe mounted on a work bench

is far more expensive than placing a double floor in the building when it is put up.

In order to have the shop well lighted, it is a good plan to put in a sky-light about 6 square feet in size. In the winter when the shop becomes more difficult to keep at a comfortable temperature, the sky-light can be boarded up, as windows always make a building cold. However, all the windows possible should be used when the weather conditions permit. Plenty of light means accuracy in machine work, and accuracy means good working models that will function as their builder intended.

The model engineer should take a great pride in his workshop and everything should be kept in good condition. After each job is finished, the tools should be put away, each in its proper place, and the bench and lathe carefully brushed off. The importance of keeping every tool in its place cannot be overestimated, as every good mechanic well realizes. If this is done, a job can be carried on quickly and the worker will not lose patience and temper in looking for a certain tool which may be obscured under a lot of dirt and junk spread about the bench. Order should be the watchword of the model engineer in regard to his shop.

CHAPTER II

LATHES AND LATHE WORK

The type of lathe to purchase for model making—Setting up the lathe—Elementary lathe work—Wood turning—Grinding tools—Knurling—Metal spinning—Turning crankshafts—Screw cutting—Internal lathe work—Attachments for a model engineer's lathe—A small lathe made portable by mounting it in a cabinet.

The lathe for model making or light experimental work need not be an expensive one. A complete outfit, comprising a practical lathe with a few tools and attachments for nearly all ordinary jobs, may be purchased for from $40.00 to $60.00, depending upon the equipment desired with it.

Money so invested is well spent, for not only does the home lathe offer opportunities for developing a fascinating and edifying hobby, but it will also provide many

Fig. 7—A model maker's lathe

a chance to turn an honest penny for its owner. The field to-day for mechanical toys, novelties, and working models is tremendous and the wise home mechanic will

25

make his work lucrative to the extent of paying for his equipment and perhaps giving him a little surplus besides. And, in this connection, let it be said that there is probably no single tool in the entire shop that develops in its owner and user such a sense of affection as the lathe.

In selecting the lathe it is well to send for the catalogs of the manufacturers, comparing the various features of

Fig. 8—Using a sawing attachment on the lathe shown in Fig. 7

practical importance as well as the prices. There are just a few *essentials* that the prospective purchaser should look out for, regardless of whether he wishes a heavy, expensive tool or a light bench lathe. One of these is the *hollow spindle* found on every modern lathe of any value whatever regardless of its price. There must be a hole clear through the live spindle (the revolving one) and the hole at the "business end" should be tapered to take a standard No. 1 Morse taper in the case of a small or moderate-sized lathe. This is important.

On the headstock (live spindle) end of the lathe there
are other features to look out for. For one thing, there
positively must be some means for taking up the inevitable wear in the spindle and bearings. The only practical method is the cone bearing, which is so simple and
effective that no honest manufacturer should do without
it even in a cheap lathe. In selecting a lathe, see whether
there is a little collar at the left hand end of the live

Fig. 9—A close-up view of the sawing attachment

spindle that can be screwed up to tighten the spindle in
its bearings.

A third point to look out for is to see that there are
regular oilcups of some description on the headstock.
Even crude ones will be better than nothing, but the really
well-designed tool, even though it be an inexpensive one,
will have a cup with some sort of a cover device to prevent chips and filings from getting into the oil receptacle
and from there to the bearing where they would cause
trouble.

The fourth point, is to see that there is a cone pulley
on the lathe. At least two steps and preferably three
should be demanded, as the only practicable method of
changing speed, and at the same time producing the corresponding change in *power delivered at the work,* is by
means of the cone pulley. The one exception to this state-

ment is found in the case of the expensive, direct-connected, motor-driven lathes. The popular idea that a variable-speed electric motor can be belted direct to the lathe pulley and a satisfactory adjustment of speed and power obtained is a fallacy, as the variations in motor speed produce a corresponding variation in power delivered at the work. The proper way to belt a motor direct to a lathe of this kind is by means of a cone pulley *on the motor* as well as on the lathe. By this means, when greater power is required at the work, the small pulley on the motor will revolve at high (proper) speed while the large pulley on the lathe will take all of the power

Fig. 10—Centering a steel rod

possible from the belt owing to the far greater tractive surface presented to the belt by the large pulley.

Other considerations are of lesser importance, but among them may be noted the size of the live spindle, which should be as large as practicable to afford stiffness, the proportions of the nose (threaded portion of spindle), which should have plenty of metal under the threads, and the size of the hole in the spindle.

The bed of the lathe should by all means be machined (milled). Some very cheap lathes are turned out for wood turning with the beds simply cleaned up on a grinder, and these are an abomination.

The tailstock (opposite the headstock) should be adjustable along the bed by means of a screw clamp readily accessible, for this adjustment is made hundreds of times during the course of a job. The feature of a quick-fed tailstock spindle is a valuable one as it adapts the lathe for drilling. In addition to the lever feed, the tailstock spindle should have also a screw feed for use in turning to hold the dead center against the work.

The purchase of attachments will rest with the individual. He will need, first of all, some tools to work with and also either a countershaft, foot power, or electric or other motor, for the drive. The countershaft is a great convenience even with direct electric motor drive, as it enables the lathe to be started and stopped almost instantly without having the momentum of the motor armature to be overcome. A drill chuck is practically essential, as is also a small universal scroll chuck. The latter will make it possible to grip metal rods, cylinders and discs in the lathe without the trouble of centering and supporting by means of dogs and clamps. However, as these are both more or less identified with rather advanced metal turning, they may be omitted until the worker has gained some experience.

The equipment supplied with the lathe, as a rule, consists of a slotted face plate (to screw on nose), an arbor with a nut to take grinder wheels, saws, etc., and also discs of metal or wood to be turned, a tee-rest to form a support for hand-turning tools, and a drill chuck to take drills from 0 to ¼ inch in size. This, with the addition of a few tools, will enable the worker to start serious work.

When the lathe comes, it will be covered with a thick, sticky grease which protects it from rust in transit. This grease must be removed with a cloth moistened in kerosene and the bright metal parts polished so that they are not sticky. Going over the lathe once a week with an oily cloth will keep all parts bright and shiny, as they should be.

When selecting the spot for the lathe, bear in mind just a few very important points. First is the light; this really should come from over the right shoulder, as it should fall directly upon the work at the point of cutting. If this cannot be arranged, the light may come toward the worker, who will then have to wear a visor to shield his eyes. The least favorable is to have the light come from over the left shoulder. It merely blinds by reflection and casts a deep shadow at the point where it is most needed. The only alternative is to place a sheet of white reflecting material on a movable arm so that the light is reflected upon the work.

The next consideration is the bench or table upon which the lathe is mounted. This bench should be stiff and rigid to resist the tendency of the lathe to vibrate, especially when running at high speed with a piece of work of unequal proportions (out of balance) between centers. If only a kitchen table is obtainable, it should be braced with crossed wires and spreaders between the legs and the feet really should be secured to the floor if possible. This latter is positively necessary if the countershaft or driving motor is on the ceiling or wall. If the motor is mounted upon or underneath the table, the feet need not be fastened to the floor. The best and most satisfactory bench is one built into the shop. It may be crude, but it must be stiff and *level;* even though ceiling, floor and walls are out of true, see that the lathe bench and all shelves, supports, etc., carrying running machinery of any kind, are quite level.

Another point is to see that the lefthand end of the lathe is not obstructed so that rods, etc., may be passed through the live spindle. See also that there is enough room in back of the lathe to permit a circular saw to be used with effect.

Set the lathe up with the largest screws that will pass through the holes in the feet, or, better still, use flat head bolts passed through holes in the bench, with the nuts and washers underneath. Make sure the lathe bed is parallel with the line shaft, or with the countershaft. Arrange all tools, attachments, etc., on ledges and shelves within easy reach and bear in mind constantly in placing these shelves that the entire bench all around the lathe becomes literally snowed under with chips and shavings when wood turning is done. It is far better to have the tool shelves slightly beyond reach of the shower of chips, as it will save many a valuable minute in cleaning up.

To line up countershaft or line shaft pulleys with those on the lathe, simply tie a plumb line to the upper shaft and slide it along until the bob comes to a standstill directly beside the lathe pulley cone. If line shaft, countershaft, and lathe have been properly mounted in parallel and perfectly level, no trouble will be had with belts running off. The final adjustment of the pulleys will make them ready for belting.

In placing belts, put the smooth side of the belt next to the pulley. The rough side will not pull nearly so well, as its uneven surface provides a myriad of air pockets which prevent traction.

To lace the belt ends, cut the belting with a sharp knife *against a tri-square,* as the ends must be square, cutting the belt ¼ inch shorter than you can stretch it by pulling the leather over the pulleys. Then punch or drill three small holes in each belt end, making sure the holes are opposite each other. Pass a piece of soft iron or brass wire, of about No. 22 gauge for a 1-inch belt, through one hole at the edge, then through the corresponding hole in the opposite end of the belt, making sure the belt is in position. Draw up the wire and pass one end through the next hole, crossing over on the rough or outside of the belt. Pass through the opposite hole and over to the third pair, crossing again on the rough side. Draw up tight and pass through once more to give a double thick lace on the side next the pulley. Cross back to the center pair of holes and pass under again. Then cross to the first pair of holes, pass under and up through and two ends of wire will be had to link together on the rough or outside surface of the belt. The smooth side, next the pulley, will have no cross-over laces to wear through after a short amount of use.

The amateur will first perform the operation known as "scraping" by the skilled worker, who looks askance upon this form of wood turning. It is, however, safe and easy in the hands of the unskilled, although it is infinitely slower, less efficient, and productive of poorer

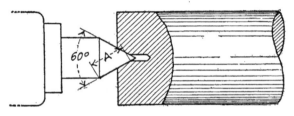

Fig. 11—Proper centering

results than the true wood turner's cut, which is more of a chisel or knife-edge cut. The latter is somewhat dangerous in the hands of the novice, as the tool is almost sure to "bite" or grab into the wood if it is not held properly.

For such tools, the use of old flat files, ground practically square on the ends, is recommended. Such a tool, when brought up to a piece of wood in the lathe, and supported upon the tee rest, which should be just as close to the work as possible, will scrape off stock at an astonishing rate. For roughing off the surplus stock, a turner's gouge is used, as a rule. This tool may be used with comparative safety by the novice if he is careful to take light cuts and to grip the tool firmly, using his forearm to hold down the long handle. The finishing cuts are taken with what is known as a skew chisel (the cutting edge is askew) but the use of this tool is not recommended unless the worker can get some instruction in its use. It will bite fiercely if not used properly, and if it does nothing worse it may spoil a piece of nearly finished work.

For metal work, the slide rest is practically an essential. Some metal turning may be done with the tool made from a flat file held on the tee rest, but for fairly good work the slide rest is necessary. This is particularly the case when a long shoulder is to be turned, for

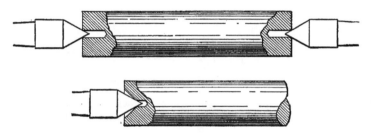

Fig. 12—Two examples of improper centering

instance. For curved profiles, the hand tool is quite satisfactory, however.

The metal turning tools that go into the tool post on the slide rest are usually of ¼ x ¼-inch carbon steel for a small lathe. There is a little curved rocker piece that goes beneath the tool in the tool post to alter the height of the cutting edge of the tool. This cutting edge should ordinarily be exactly on a line with the center of the work. There are modifications of this rule, particularly in the case of work of large diameter, but in facing or shouldering small rods, for instance, the tool should be set to the center.

There are just a few cardinal principles to be mastered by the worker in metals that are essential even to the most mediocre work. These points will be considered. The first thing to learn is what is meant by *rake* and *clearance*. These two terms mean volumes in the use of metal working tools.

The rake of a tool is the angle of the top or cutting surface to the work being cut. If the surface of the tool slopes *toward* the work, or downward in the direction of the work, the rake is *negative;* if it slopes *away* from the work, the rake is positive. To keep these in mind, remember that the *positive rake* is the one that makes a *barb or hook* of the tool to catch underneath the chip and pull it away. The negative rake *scrapes* the surface away instead of catching underneath and tearing it or pulling it.

Fig. 13—Lathe centers should be kept as shown at the left. Accurate work cannot be done with a tool such as that shown at the right

Now there is something in the nature of metals which we do not fully understand but which experience has taught us makes it necessary for us to use one kind of rake with certain metals and the opposite with others. For instance, *brass invariably demands a negative rake.* If one attempts the slightest operation upon brass with a tool having a positive rake, the tool is almost sure to bite into the work, spoiling it and perhaps break the tool. Be sure to have the top surface of the tool ground so that it slopes downward toward the work when cutting brass. Steel, on the other hand, demands a decided positive rake.

It is not necessary to regrind the entire cutting surface of the tool when changing from one to the other. If the tool has normally a positive rake for steel, one needs merely to take the tip of this rake off so that just the cutting edge has a negative rake when brass is to be cut.

In turning from a casting, take a cut sufficiently deep the first time to get quite beneath the scale or hard crust on the casting. Be sure this crust is cut right through, as, if it scrapes the tool at all, the cutting power of the latter will be destroyed and it must be reground.

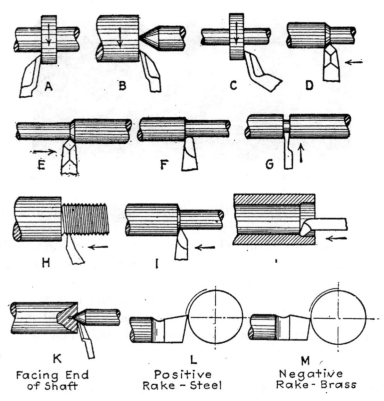

Fig. 14—A. Using a left-hand side tool. B. Using a right-hand side tool. C. A right-hand bent tool. D. A right-hand diamond point tool. E. Left-hand diamond point tool. F. A round nose tool. G. A cutting-off or parting tool. H. Bent threading tool. I. Roughing tool. J. Boring tool. K. Facing the end of a shaft with a right-hand facing tool. L. A tool ground with a positive rake for cutting steel. M. A tool ground with a negative rake for cutting brass.

The *clearance* of a tool is the separation or, as its name implies, the clearance between the body or supporting portion of the tool and the work being cut. The illustration showing the various angles for cutting tools illustrates this clearly. All tools must have clearance to prevent the unused portion of the tool from scraping the

Fig. 15—Thread cutting on a small lathe with a die and stock

work and preventing the cutting edge from doing its duty.

The larger lathe shown in Fig. 17 is very suitable for general amateur work and model making, although it is much more expensive and requires greater power to drive it. Unlike a small lathe, this machine is capable of doing many special jobs that could not be done successfully on a smaller machine. The smaller the number of machines in a shop, the more desirable it is to know how to use the lathe for work not ordinarily done on it. A drill press is the proper machine tool to employ in boring holes. But, if we have no drill press, perhaps we may be glad enough to rig up the lathe for the work, even though it will not do the job quite so economically.

The lathe may be used for putting on a checked edge or milled edge or the like on the heads of thumb screws,

thumb nuts and similar articles. Knurling may be done
on brass or steel work by using the proper tools and
speed. For what may be called *hand knurling,* we use
the small steps on the cone pulley, if the lathe is driven
by one; or, if electrically driven, we employ whatever
means we may have to get speed. Brass may be knurled
properly with a surface speed in the neighborhood of
300 feet per minute; steel, at 200 feet. These are high
speeds and may not be obtainable without making special
arrangements. If not obtainable at all, then we may use
the best speed possible. The speeds mentioned refer
to the rate at which the rotating edge passes a given
fixed point. A 1-inch brass screw head would have to
turn round 1,145 times in a minute in order to have a
peripheral speed of 300 feet per minute. For a steel

Fig. 16—Turning down a small steel rod

screw head of the same size, the 200 feet per minute would
demand 763 turns in a minute of time. Where the tool
is mechanically controlled, we may reduce these speeds
to very ordinary ones; say, 60 feet per minute for brass,
and 40 feet for steel.

A special knurling tool is used, which the operator
may buy or make for himself. The type used in hand
knurling consists of a knurling wheel mounted in a metal

holder at one of its ends. At the other end of the metal holder, a wooden handle may be arranged for the convenience of the operator. The knurling tool used in machine knurling may consist of two knurling wheels mounted in a small bracket, which in turn is pivoted on a holder adapted to be held in the tool post.

With hand knurling it will be advantageous, not to say necessary, to have a rest mounted in front of the

Fig. 17—A large model engineers' lathe

work. The tool is held in the right hand and is laid on the rest, where it is controlled by the left hand. Matters are managed so as to bring the knurling wheel up against the work from below and toward the front.

With machine knurling, we secure the tool in the tool post, setting it so as to bring the knurling wheels up against the under part of the work, but with the nearer wheel well forward of the axis. The position of the wheel, relative to the work, is substantially the same whether we control the tool by hand or by means of the tool post. In using the two-wheeled tool we manage so that both wheels shall press equally against the work at all points.

In beginning, whether with the hand controlled tool or not, we oil the work and the knurling wheel or wheels. In machine knurling, with the pair of wheels, there is more or less danger of spoiling the work by making different indentations with the two wheels. To govern this matter, things may be tested before putting on power. We press the tool by hand up against the work, using

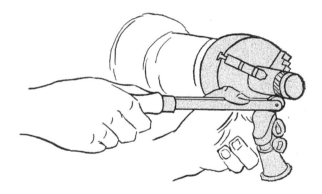

Fig. 18—Knurling with a hand tool

considerable force, and work the lathe belt by hand. When we have started the knurling *all round* and are sure everything is right, we may go ahead and use power drive.

A crankshaft will have one or more cranks. Sometimes these are set differently on the shaft. The turning of the wrist pins is a particular job, requiring exact and painstaking attention. The difficulty is, however, not so much in the actual cutting operation as in the preliminary work of getting the crankshaft properly on the lathe. Naturally, it is held by the centers in the head and tail stocks. The line connecting the points of these centers must be in line with the axis of the wrist pin and both must be exactly parallel with the axis of the shaft itself. When we go to put the work on the lathe centers we will probably find that not only have we no center holes for the centers but no places to put such holes. This difficulty may be overcome by putting temporary arms on the ends of the crankshaft and boring center holes in

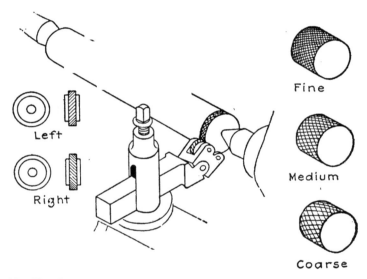

Fine

Left

Medium

Right

Coarse

Fig. 19—Knurling with a machine tool. Examples of knurling are shown at the right

these pieces. Each of these arms corresponds to one of the arms of the crank whose wrist pin it is proposed to turn. It may be secured to the shaft by arranging a suitable hole in the arm to receive the shaft and then effecting a grip by means of a set screw arranged in the arm. If the crankshaft has cranks set at angles to one another, then we may use a kind of elbow instead of a simple straight arm. Each part of the elbow will correspond to a crank and the shaft will pass into a hole placed at the angle of the elbow. The center holes for use in machining the wrist pins are bored in the outer ends of the arms. If we have only single arms to deal with, then the location of the center holes will not be so difficult a job as otherwise. We first ascertain the exact length of the throw of the crank. This is the distance apart of the axes of the wrist pin and the shaft. If the two flat faces of the arm are planned parallel, we will be in good position to locate the central points of the two holes, one for the shaft and the other for the center hole. The problem now is to find if everything is all right. If the set screws are arranged precisely the same

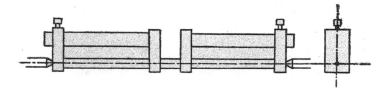

Fig. 20—How a small crankshaft is mounted between lathe centers

in both arms and are, in fact, set in line with a plane through the axes of the two holes in the arm, then the tightening up of these screws should put the two center holes at exactly equal distances from the axis of the shaft. Tighten up moderately and check with a surface gauge. The work may be rotated by operating the lathe by pulling on the belt by hand. Note at the same time whether there is going to be stock all around for the wrist pin. The testing of this matter may be done by putting a sharp-pointed tool into the tool post and bringing it up near the wrist pin. As the lathe is slowly worked to rotate the crankshaft, we can note the space between tool and wrist pin.

The lathe may be used for spinning, provided two conditions may be satisfactorily met. One of these is substantial construction; and the other is speed. Spinning requires the speed in the spindle of the lathe itself ranging from 1,800 to 2,500 R.P.M. ·Whatever arrangements are made to get the speed, they should be substantial, as it is necessary to rotate the spindle, the work, and accessories used to secure the work to the spindle,

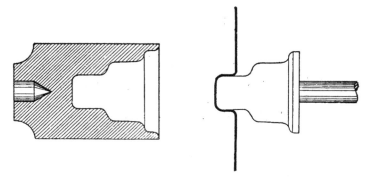

Fig. 21—Wooden follower for metal spinning

as a face-plate, etc., and in addition overcome the resistance to the spinning operation. With a countershaft connected up so as to run at 450 to 600 R.P.M., we have only to effect a drive of the spindle by pulleys in the ratio of 4 : 1 to get from 1,800 to 2,400 R.P.M., no account being taken of the slippage of the belt.

There are three or four differences in the lathe as used for metal spinning from the lathe as used for ordinary metal turning. These differences affect the headstock, face-plate, the tool rest and the tail center. The "dog" is not employed. The ordinary center-screw, faceplate or outside screw face-plate used in wood turning is screwed onto the headstock spindle. A block of hard wood, e.g., hard maple, is screwed onto the face-plate and then turned to the form corresponding to the shape that is to be assumed by the work at its first stage. The tail center is used to hold against it the circular disc of sheet metal which constitutes the blank for the spinning. Friction is depended upon to hold the disc in place, until actual spinning begins. The rest is not a difficult thing to make. The tail center may be purchased and the tools used are not especially hard to make. The metals in common use for spinning are copper, white metal, brass, zinc, and aluminum. A great range of articles can be made. Many articles that can be made by stamping can also be made by spinning. Some articles can probably be made better by spinning them. The tool is used as a lever, one end being pressed against the work and the hand supplying the power at the other. A movable pin set up on the rest provides the fulcrum of the lever. The direction of motion of the lever varies from a vertical

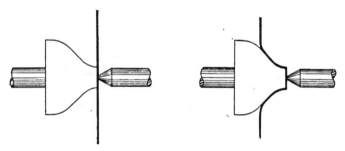

Fig. 22—How a piece of metal is spun over the form

plane at the beginning of operations to a horizontal plane at the completion of the spinning process. As the metal is gradually pressed over the form, the pin is moved toward the headstock of the lathe by changing it from one hole to another. The ordinary speed lathe, if of rather robust construction, is more suitable for this work than the engine lathe. It is a question of the speed of the spindle. If that can be obtained with the ordinary substantial engine lathe, then the problem is solved.

Internal work on the metal working lathe is largely done by means of the boring bar. By mounting the bar on the lathe centers and providing for its rotation by one or more dogs, we have a considerable part of our preparatory rigging. The tool or tools are held by the bar, a slot cut through the latter providing a means for holding the shank of the cutting tool. The tool is rotated, but is not, in the method set forth, fed to the work either parallel with the axis of the lathe or perpendicular to it. These two movements constitute, in external work, the feeding of the tool. In the present style of internal work, the tool has only one motion and that is rotatory. The

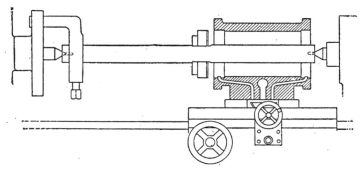

Fig. 23—How a boring bar is employed to bore out a steam engine cylinder

longitudinal and transverse feed is accomplished by carrying the work along the lathe axis and shifting it transversely. The work is secured, very firmly, to the carriage or rather to the lower portion of the compound rest. The longitudinal movement of the work back and forth can be provided for by putting the carriage into mesh with the lead screw. The transverse movement of the work is secured by operating the bottom part of the rest back and forth squarely across the carriage. The tool in the bar may be held in place by one or more wedges in the slot. A great deal of work may be done by this rig.

As this type of boring bar is quite suitable for boring out cylinders for gasolene engines and the like, a few more remarks may be justified. To test whether, at the beginning, the bar and the work are central, we may replace the cutting tool in the bar by a piece of wire. We run the carriage back and forth and rotate the bar, all by hand, and thus see whether work and bar agree. In case of disagreement, it ought not to be the bar that is at fault, since it is set between the lathe centers. But the work may not be set properly on the carriage. In

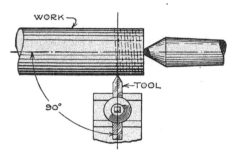

Fig. 24—How a thread-cutting tool should be mounted

making the finishing cut, we should use an almost microscopic feed in order not to spring the bar. Further, the final cut should be taken at one operation without stopping. One reason for this is that stopping means cooling off, and cooling off means shrinkage of tool and expansion of the hole being cut.

Probably the most important use of the lathe is that of screw cutting. There are several kinds of screw threads in use. In America, the 60-degree V-thread and the United States Standard or Sellers thread are employed, and in England the Whitworth. Besides these are what may be called the square thread and the Briggs pipe thread. In all, the type is defined by specifying details relative to an axial section. The 60-degree V-thread is one whose axial section is a triangle, the angles of which are all 60 degrees. The U. S. section is similar, only the tops of threads and bottoms of grooves are flattened. The Whitworth thread is based on an isosceles triangle with the vertical angle equal to 55 degrees. The tops and bottoms of the grooves are rounded. The Briggs thread increases or decreases in diameter in passing from thread to thread. The thread is otherwise something similar to the Whitworth, only the triangle is equiangular and the roundings are on a much smaller radius of curvature.   •

The cutting edges of the cutting tool which may be used to form any of the usual non-tapered threads is properly shaped to the exact form and size of the axial

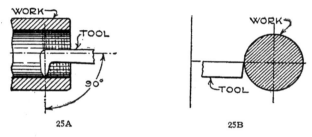

25A   25B

Fig. 25A—Setting the tool for internal thread cutting
Fig. 25B—The thread-cutting tool should be set at the exact center of
the work

section desired at the finish. That is, the flat horizontal top surface of the nose of the tool should have the required form and size. The nose should also be so shaped that as the top is ground down from time to time for the purpose of sharpening the tool, the form and size of the thread section will be maintained. This shaping of the nose may be very accurately done at the beginning, before the tool goes into action. A gauge may be made of a piece of thin sheet steel by cutting a notch of the precise size and form of the thread wanted. When grinding the nose of the tool on the front and on the sides, this gauge may be used to test the work from time to time. The top surface of the nose is properly made flat and parallel to the top and bottom surfaces of the shank. The object in view is to present to the work a cutting edge that is horizontal. If the parallelism is not provided, then we may expect the tool to cut a groove too narrow or too wide. If exact results are wanted, too much care can hardly be given to the grinding of the nose. Once ground to size and form, the regrinding should be comparatively simple, as all it is necessary to do is to maintain flatness

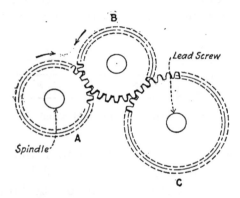

Fig. 26—The arrangement of the gear wheels at the end of a screw-cutting lathe

and parallelism with the shank. The size and form of the section will then be right.

When the tool is set, it is very important that it be horizontal and that the top surface of the nose be at the exact level of the axis of the work. If this requirement is disregarded, then we may expect the thread to be wrong. Thus, if the top of the thread is to have a 60-degree angle and we set the tool too high or too low, then we will not get 60 degrees but something different. If the question is asked, "How is one to make sure he has the cutting edge at the exact level of the axis?" the following answer may be made: Put the centers in the head-stock and tail-stock of the lathe. Then bring the tool up close to each and note whether it is in agreement with the level of the points of the centers. There is still another requirement. The axis of the cutting edge and the axis of the shank should be exactly parallel. When the tool is set, these axes must be perpendicular to the axis of the work.

In order to cut the winding groove on the work, it is necessary that the tool shall move along parallel to the axis of the work while the latter is rotating. In fact, there must be a very exact correspondence between the

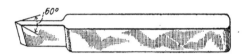

Fig. 27—How a thread-cutting tool should be ground

forward or backward shift of the tool and the rotation of the work. This shifting of the tool is ordinarily secured by means of the lead screw of the lathe. We put the carriage which supports the tool and tool post into the control of the lead screw. When the lead screw turns around once, the carriage and tool will be shifted exactly the amount of the pitch of the lead screw. That is to say, for example, if the lead screw has 6 threads to the inch, then the pitch will be exactly ⅙ inch. Suppose, now, that when the work turns around once, the lead screw also turns around once. Then, we should have the tool advancing or receding ⅙ inch with every turn of the work. In fact, we should cut a thread of exactly the same pitch as that of the lead screw. But, if the work rotates faster than the lead screw, the tool will be shifted too slowly to cut a thread of the same pitch. We should get a thread of a somewhat different pitch. Similarly, if the work rotates more slowly than the lead screw, the tool will shift too rapidly to cut a thread of 6 convolutions to the inch. We will get a coarser pitch, which means a smaller number of threads to the inch. It is possible to regulate the rotation of the lead screw relatively to the spindle of the lathe and get just about any pitch on the work that we desire. If we want 12 threads to the inch, then we must make the lead screw turn half as fast as the work or spindle. If we want 3 threads to the inch, then we must adjust the lead screw to rotate twice as rapidly as the work.

The lead screw is usually driven by the spindle through gear wheels. It is not especially difficult to learn

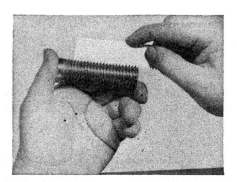

Fig. 28—Testing threads with a small pocket gauge

how the gears control the pitch of the screw thread we cut, and what gears to use in order to get a certain pitch that may be desired. Let A, Fig. 26, be a gear wheel on the spindle; C one on the lead screw; and B, an intermediate gear. First, consider B. It serves to keep the direction of rotation alike between spindle and lead screw. If the spindle gear A rotates *with* the hands of a clock, then the lead screw gear C will also rotate *with* the hands; and *vice versa*. Second, the number of teeth on B plays no part in the relative speeds of A and C. For example, suppose A and C have, respectively, 32 and 28 teeth, then a complete rotation of A will produce

1⅐ rotations of C. It will make no difference whether B has 10 or 40 teeth. Consequently, any figuring we have to do will not need to take the intermediate gear into consideration. It simply serves to keep the rotation directions of A and C the same.

Now suppose we want to cut 10 threads to the inch and that our lead screw has 6 threads to the inch. What must be done is to select proper gears for spindle and lead screw to give 10 turns of the spindle while we get 6 turns of the lead screw. The gear A will be the smaller one. Further, the two gears must have the numbers of their teeth such that these numbers will be in the ratio 6 : 10. If one has 18 teeth and the other 30, that will cover the case. Or, one could have 24 and the other 40. In fact, it doesn't matter what the numbers themselves are, just so that we have the right ratio, 6 : 10. Then we put the smaller one on the spindle and the larger one on the lead screw. It may be helpful to recollect that the more slowly the lead screw turns, the finer the thread will be.

Take another case. Suppose we want to cut a coarse thread 5 turns to the inch. This is coarser than the thread on the lead screw itself. Consequently, we want the lead screw to turn more rapidly than the spindle or the work. This means that the smaller gear must go on the lead screw. All that remains to do is to select gears for lead screw and spindle that are in the ratio 5 : 6, put the bigger one on the spindle and the smaller one on the lead screw; and select an intermediate gear to make it possible for the one to drive the other. Gears having 10 and 12 teeth, 15 and 18, 20 and 24, 25 and 30, etc., are all suitable.

If the lead screw has a right hand thread, an intermediate gear, or some equivalent, will be needed when we want to cut a right hand thread. However, a right hand lead screw and no intermediate gear, or else two of them, will cut a left hand thread. A right hand thread is cut by advancing from right to left, and a left hand thread by advancing in the opposite direction.

The work may be held on the lathe between centers or it may be held by a chuck. In general, work carried between centers may be cut more accurately than if the chuck carries it. This is due largely or entirely to the double support. It is a good rule when working with a chuck on the head-stock never to take the work. out between the beginning and end of all turning operations. This would apply to cutting screw threads. In fact, it would probably be quite difficult, if not impossible, to cut a reasonably perfect thread, if the work is disturbed when half done.

Before cutting a thread between centers, it is advisable to make sure that the centers themselves are right. The point of a center should be exactly on the axis of the center. The levels of the two centers must be exactly alike. To test this, the screw cutting tool may be set in the tool post and the carriage run to one center and then to the other for the purpose of setting the tool for height at one and of testing the other center for agreement with this level. It may not be out of place here to say a few words about centers. The work turns about the tailstock center. It is advisable, then, to prepare the hole in the work at the tail-stock and so that the point of the center and the metal of the work will not be in actual contact. This may be done by first preparing a conical hole to fit the center and then counterboring it at the bottom with a small drill. This drill hole, if deep enough, tends to prevent damage to the center point either by wear or by friction. It is well to counterbore the other end of the work also. Fig. 11 will make this very clear.

If an interior thread is to be cut, we will naturally have to use a tool somewhat different from the plain straight tool for cutting exterior threads. A suitable tool for a considerable range of work is one with a right-angle bend in it near the nose end. We are then able to move the tool back and forth in the hole. Aside from the bend, the tool may be precisely the same as the one already described. It is very essential that the flat top of the nose shall be set at the exact level of the axis of the work and that the axis of the flat top be exactly at right angles with the axis of the shank. This latter requirement is the one, perhaps, that will make the most difficulty. It will be well to have a substantial shank so that the stress of cutting will be well resisted. This resistance may be increased, also, by shortening the distance from the bend to the point where the tool holder grasps the shank.

Whether we cut an interior or an exterior thread, the tool will naturally wear. This wear should be confined to the edge of the top. To sharpen the tool and perhaps better its shape, it is reground on the top surface of the nose. The final surface should be exactly parallel with the top and bottom surfaces of the shank. Naturally, a reground tool will not have its cutting edge at the proper level, but below it, unless we take special measures for correcting the level. This we may often do by simply putting a strip of thin sheet metal beneath the shank in the tool holder.

A little consideration will perhaps convince the reader that when we screw one thread into another, it is not so important that the top of one thread shall touch the bottom of the other as that the body of one thread shall fit snugly into the groove of the other. In fact, we may have the case where the top of neither thread reaches quite to the bottom of the other and still not have any noticeable defect. If the top edge of a screw thread is somewhat worn, we may not be able to tell its real diameter by measuring the over-all diameter. It will, sometimes at least, be best to rely on what is called the *pitch diameter*. This is the average between the diameter measured from top to top of the thread and the diameter from bottom to bottom of the groove. It is really the distance from half way between top and bottom on one side to half way between top and bottom on the other. This is a matter of some importance for the reason that it may be necessary sometimes to take off the top edges of 60-degree V-threads to prevent trouble when screwing into each other. Indeed it is good practice both with the sharp V-thread and the U. S. Standard to cut off a trifle at the top of the thread, provided the thread is not to be case hardened. There is the advantage that a close fit can then probably be better made than otherwise. The reason for thinking so is this: Very slight differences between the two threads will then have an opportunity for rectification, the metal having a chance to flow into the open space. That is, the two threads can force each other a trifle.

Where a good deal of work of one size has to be done, it may be well to use taps and dies. There are taps for use on a power-driven machine and taps for hand use. Similarly, with dies. In general, accurate thread cutting should not be attempted with hand-operated tools. It is, for one thing, too difficult to be sure that the axis of the tool and the axis of the work are exactly in line during the operation.

We now come to the question—How are we to determine whether our threads are right or not? We may try one with the other. But this is by no means reliable. Two threads that properly fit together bear against each other throughout. Thus a nut and a bolt when the one is screwed onto the other should so fit round and round the thread that a strain tending to pull the nut off would be resisted by all the convolutions in engagement and not by one or two only. It is possible, however, for one

to fit a nut and screw together without being able to tell
whether there is bearing of thread against thread
throughout. If the thread in the nut has a pitch a trifle
longer than that on the screw, the nut might seem to
have a proper fit when tried merely by screwing it on,

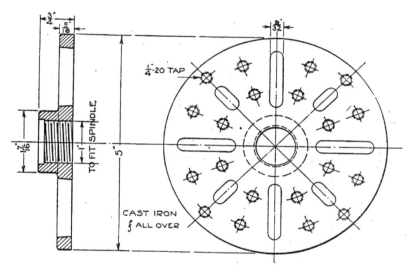

Fig. 29—A face plate for a small lathe

because of the contact of threads at the two ends. We
could, if the screw thread goes further inward, test the
matter of unequal pitch by trying to screw the nut in
further. If the resistance is strong—stronger than when
we were simply putting the nut on—then we probably
have a case of inequality in pitch.

However, there is a very simple instrument by means
of which we can determine whether the pitch of the screw
agrees with the standard required or not. This is a
short strip of thin metal on one edge of which teeth
somewhat like those of a saw have been cut. These are
really the axial sections of the grooves desired on the
screw. When held so as to fit into the valleys on top of

a horizontally held screw, a very minute error in pitch can be readily detected, especially if the light is back of the device. The use of this tool is shown in Fig. 28. It is understood that a good gauge of this type will enable a beginner to detect a pitch error of only 0.005 inch. These gauges are not to be confused with the ordinary pitch-gauge used simply to tell whether the screw has 12 or 13 threads to the inch or the like. Precision pitch gauges are tested to a high degree of accuracy—one concern, at least, claiming an accuracy of 0.0001 inch. They are not so applicable to interior threads. But the gauge may be applied to the tap, if one has been used to make the interior thread. If the thread has been cut on the lathe, then we may have to depend a good deal upon the

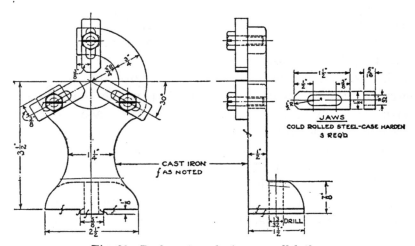

Fig. 30—Back rest made for a small lathe

fact that we used the *same* combination of gears for screw and hole.

Many amateur mechanics have small lathes in their workshops which are very limited in their application, owing to the fact that they are not provided with many of the attachments that are found on larger machines.

The following paragraphs· describe a few useful appliances that may be easily made and attached to the original small lathe and will increase its usefulness considerably.

The drawing, Fig. 29, shows a face plate which is made up from cast iron. The slots are cut and the center hole is left smaller than the diameter of the nose of the spindle so it can be bored out later and threaded. Quarter-inch holes are then drilled and tapped as shown in the sketch. It will be necessary to locate these holes systematically about the surface of the plate and the more holes it contains the easier it will be to clamp odd shaped work to it, as in boring, etc. The center hole of the face plate must be drilled and threaded to fit the lathe spindle and this will, of course, depend upon the size of the spindle on the lathe the plate is to be attached to. The mechanic who has a very small lathe will find it quite impossible to do turning on the face plate with his own machine and will therefore find it necessary to take the work to a local shop which is equipped with a screw-cutting lathe. While the face plate is fastened on the same lathe on which it is being threaded, the back hub must be faced off so it will run true with the thread. It is best to turn the face plate off when it is in position on the small lathe on which it is to be used to make sure that it is running true. After the rough turning is done, the face plate can be polished up and finished.

Another useful attachment is the back rest, shown in Fig. 30, and this is of very simple design. The standard is made of cast iron and the bottom, which fits in the bed, can be either filed or milled. While it would be very practical to have the job done on a milling machine, if the mechanic does not have access to such a machine he can file the casting if a little patience and care is exercised. The jaws of the back rest·are made of machinery steel and the slots can be cut in them by drilling holes down the center. The superfluous metal can be filed out to finish the slots. After this casting is finished, it is advisable to have it case-hardened so it will resist wear. When it is fastened to the bed of the lathe, the bolts on the tee rest can be used. One side of the head is cut away and this keeps it from turning. A regular nut is used for tightening it on the bottom. The screws for the slots of the jaws are provided with a hexagonal head

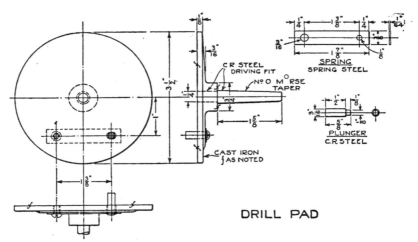

DRILL PAD

Fig. 31—An easily made drill pad for a model maker's lathe

which permits the use of a wrench to tighten them. The heads of the screws should also have a slot in them so that a screw driver can be used for preliminary tightening.

A small drill pad is illustrated which can be easily made for the tail stock of the lathe and which will enable the mechanic to do drilling operations on his machine. The pad proper is made of cast iron and the workman can easily make up a pattern of this and have it cast at a local foundry, as the pattern can be turned out with very little trouble on a small lathe. The casting is provided with a No. 0 Morse Taper. After the shank is fastened to the pad it is again put between centers and the front of the pad is faced off and finished. This will insure an accurate hole when the pad is being used. When the pad is completely finished, it may be well to enamel the back of it, as this surface is not used.

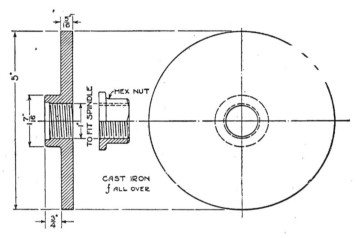

Fig. 32—Grinding disc made to attach to a small lathe spindle

The small plunger will be found very useful in preventing light pieces from turning when they are being drilled. In case some large plate surface is being drilled, the spring on the plunger will allow it to come back and permit the work to lie flat on the pad. Care should be taken to see that the work used against the plunger is not too heavy, as this would force the spindle considerably, as the only thing that would keep it from turning is a small screw.

A very convenient tool is shown in Fig. 33. This is a lead hammer which is very useful in knocking work in and out of the arbors. The nose of the hammer is very soft and yet sufficiently heavy to give it the proper momentum. The handle of the device can be used as a ram in the head stock spindle to drive out the various attachments that fit in this member.

In constructing the tools outlined in the above paragraphs, the mechanic should experience no trouble in obtaining the castings, as the patterns can be easily produced in the workshop. The face-plate and drill pad can be turned out on the lathe, and the back rest will have to be built up.

The small grinding disc shown in Fig. 32 can be very easily made and is an extremely useful tool when attached to the spindle of the small lathe, as many grinding and surfacing operations can then be performed which would otherwise be impossible. The disc is cast by means of the same pattern that was used in the pro-

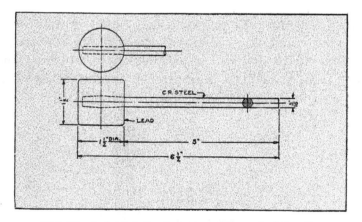

Fig. 33—A lead lathe hammer with a steel handle

duction of the face plate, previously described. No holes are drilled in the plate, however, as its surface must be machined and finished smooth. The hub is threaded to fit the spindle of the lathe and a hexagonal nut is also made to fit the spindle.

After the plate is made, it should be mounted on the spindle and, with the lathe running at high speed, it should be given a good polishing with fine emery cloth. The abrasive paper or cloth is held to its surface by means of beeswax.

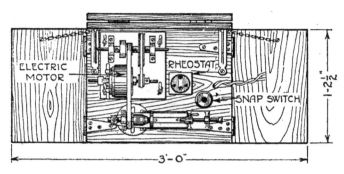

Fig. 34—General arrangement of the lathe and motor within the cabinet

In doing this, the grinding disc is set in motion at a high speed and the beeswax is pressed to the surface. The heat of friction will cause the beeswax to melt and a thin film of it will be deposited upon the polished sur-

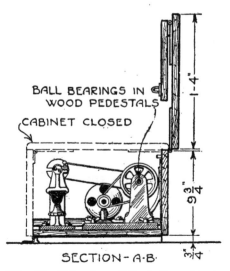

Fig. 35—Side view of the lathe cabinet

face of the disc. After this film is deposited, a disc of abrasive paper is then held tightly against the surface while it is revolving.

For the model engineer who lives in a small flat, with little available space for a workshop, the outfit described in the following paragraphs will be of great interest. It

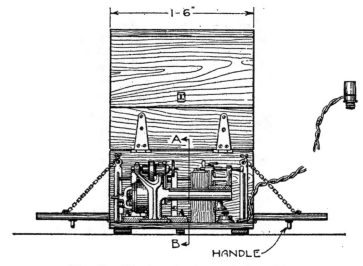

Fig. 36—Showing the front of the cabinet

is a description of a portable lathe cabinet that may be kept in the living room when not in use, and when put on the kitchen table, opened and connected to the electric light socket, forms a complete motor-driven lathe equipment capable of wood turning, metal turning, polishing, grinding, drilling, and numerous other lathe operations. The capacity of the lathe is small and the work that may be done on it is somewhat limited, though sufficient for model making.

The cabinet is made of cedar, ¾ inch thick, the boards having been secured by tearing apart an old cedar chest that had been discarded. This made it possible to provide an exterior shellac finish that presents a good appearance when the cabinet is closed. Iron strap hinges are used for hinging the end, top, and front pieces. Brass handles are provided at the ends to facilitate handling it, and incidentally, they add to the general

Fig. 37—The lathe cabinet open, ready for work

appearance. Brass dowel pins are used to hold the ends of the cabinet in perfect alignment with the top when it is closed. These are not shown on the drawing. Iron brackets are used to stiffen the back. It will be found desirable to provide four small rubber feet to prevent marring the table upon which it is placed when in use.

The lathe used is a "Goodell *P*ratt," No. 29. It has a swing of 5 inches and the extreme distance between centers is 3½ inches. The appliances used with it are as follows:

Face plate.	Saw arbor.
Drill chuck.	Grinding wheel.
Three-jaw scroll chuck.	Buffing wheel.

The cone pulley has two steps for a ¾-inch flat belt. The speeds of the lathe spindle range from 1,800 to 2,700 R.P.M. The motor is a small one of the universal type. This makes possible the use of alternating or direct current at 110 volts and develops ¹⁄₂₀th H.P.

The maximum speed of the motor is approximately 1,200 R.P.M. When first placed in the cabinet, the high speed at which it operates caused it to be noisy and it was found necessary to mount it on rubber to reduce this noise.

Fig. 38—The cabinet closed and locked

The countershaft is ⅜ inches in diameter and has ball-bearings mounted in wood pedestals. The ball-bearings are of a standard type and fit into counter-bored holes in the pedestals. The hubs of the pulleys on the

Fig. 39—View of the open cabinet from the end

shaft fit against the ball-bearings, thus holding them in place.

The pulleys are of standard manufacture, there being a grooved pulley 4 inches in diameter which is connected to the 1-inch motor pulley with a round leather belt, and two ¾-inch face pulleys 1½ inches and 2 inches in diameter from which the lathe spindle is driven.

The countershaft and motor are mounted on a wood base which is bolted to the base board of the cabinet. The bolt holes are slotted, thus making it possible to take

Fig. 40—Showing the driving motor

up slack in the lathe belt by moving the countershaft and the motor.

The rheostat is just behind the lathe where it can be reached conveniently by the operator. Near it is a snap switch which controls the entire current supply to the motor. Either this or the rheostat, or both, may be used as the operator wishes.

If the builder desires, he can arrange a small rack on the back of the cabinet to hold various tools and attachments for the lathe. If such a rack is made, it will be necessary to so design it that the tools will be held in place while the cabinet is being taken from place to place.

CHAPTER III

Marking work for drilling—How to sharpen drills for various metals—
Speed of drill for different work—Description of twist drills and
names of parts—Using the V-Block.

Drilling generally forms an essential operation in the construction of anything the model engineer makes, and knowing how to drill accurately and properly is a distinct asset that every amateur mechanic should avail himself of.

In the following paragraphs will be found a short but practical treatise on the subject of drilling which has been prepared for that class of readers who have never had the opportunity of becoming learned in general machine shop practice.

Unless the holes are to be drilled promiscuously, measuring and marking constitute the first operation in drilling any object. As a means of illustrating, we will assume that we have a brass plate 3 inches square and ¼ inch thick to be drilled with holes of various sizes. With the exception of the method of sharpening the drills, the drilling of a piece of brass is no different than the drilling of any other metal.

The tools necessary for marking are a rule, a pair of dividers, a center punch and a scribe. The scribe, which is merely a sharp-pointed piece of steel used to scratch marks on metallic surfaces, can be made from an old round file ground down on a wheel. A center punch can also be made in the same way. The rule is a steel one of the machinists' type and the dividers need not be larger than four-inch.

65

The brass plate is to be drilled as shown in Fig. 41. The first operation will be that of finding the exact center, and this can easily be done by scratching two lines as shown in Fig. 42. (Do not scratch the lines too deeply, as they will have to be papered off when the drilling is done.) At the point where the lines intersect, a small indentation is made with the center punch, as this is the exact center and all future measurements will be made from this. We will now measure for the holes in the corners. As they are 1½ inches from the center, we will open the dividers to 1½ inches, and with one point

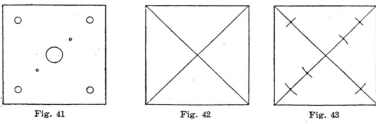

Fig. 41   Fig. 42   Fig. 43

Fig. 41—How the plate is to be drilled
Fig. 42—Finding the exact center
Fig. 43—Locating the holes

in the center, scratch a small arc in each corner of the plate so it crosses the line we first drew. This is shown in Fig. 43. With the dividers open to ¾ inch and the point in the center, two arcs are marked as shown for the two small holes which are to be ¾ inch from the center. At each point where the arcs intersect the two original lines, make a small indentation with the center punch.

With the dividers open ⅜ inch, scratch a circle in the center of the plate and within the circle draw another one, about half the size of the first one. Also scratch a small circle at each corner and for the two small holes just off the center. These circles are an aid in drilling, and their use will be described later.

Before describing the actual drilling of the brass plate, a few lines will be devoted to the twist drill and how to use and sharpen it for different classes of work.

First, let it be known to every amateur mechanic that it is absolutely impossible to drill accurately unless the drill has been sharpened properly—with mechanical exactness. In order to sharpen a drill in the proper way, an elementary understanding of its working principles is essential.

Fig. 44 shows an ordinary twist drill together with the names of its various parts. It will be noted that there is a pronounced clearance between the cutting edge A and the back edge B. Both cutting edges of a drill should be at exactly the same angle and the clearances on each side should also be as nearly equal as it is possible to make them. In sharpening a drill, the angle of the lip clearance must be left to the judgment of the mechanic, and care should be taken that it is not too great, as this will cause the drill to bite too greedily. Equally defective is a drill without enough clearance between the cutting edge and the back edge, as it will heat up excessively and also cause the flute edges to wear rapidly, thereby throwing the drill out of caliber. To those who are not experienced in grinding drills, the writer would suggest studying the clearance on new drills of various sizes. This will be found to be very helpful. In grinding the small drills (Nos. 60 to 80), care need not be taken in rounding off the clearance, as a flat clearance will suffice. Small drills should be ground on wheels of fine grit.

If the clearance on each side of a drill is not equal, it is impossible to drill accurately with it, as it will have a tendency to revolve eccentrically, owing to unequal pressure, and thereby produce a hole considerably larger than the gauge of the drill.

When a drill is to be used for drilling brass or cast

iron, it should not be sharpened in the ordinary manner. The lip of the drill should be ground off as shown in Fig. 47. Aside from the advantage of cutting faster, this prevents the drill from worming its way just as it breaks through at the end of the bore.

Assuming that the drills are accurately and properly sharpened for the drilling of brass, we will now describe

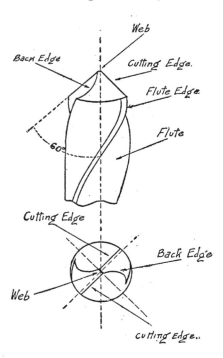

Fig. 44—A twist drill with the names of the various parts

the procedure in boring the holes in the brass plate. The large center hole shall be the first one drilled. It would be very bad practice to start drilling this hole with a ½-inch drill, as the web of such a drill is so broad that it is a very difficult matter to accurately "center" it. The only way to overcome this disadvantage is to start the hole with a drill of smaller gauge—in this case a

⅛-inch drill will do nicely. It is at this point that the circles scratched on the plate come into use. Place the ⅛-inch drill in the chuck, start the press and bring the

Fig. 45—Showing the use of a ''V-Block''

spindle down until the drill touches the center dot. Permit the drill to go just far enough to drill the dot off, then raise the spindle, and, by means of the small circle scratched on the plate, see if the tiny indentation made is exactly in the center, using the circle as a guide. If it is properly centered, drill just a little further (do not permit the point of the drill to go very far below the surface) and follow out the same operation on each corner. When the ½-inch hole in the center is drilled, go cautiously until the drill is centered accurately, and do

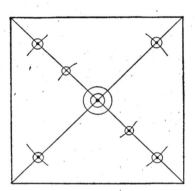

Fig. 46—Plate showing the holes started after the circles are made

not bore right through the plate without raising the spindle several times to see that the drill is in the exact center. Fig. 46 will make this clear.

Those amateur mechanics who have tried to drill a transverse hole in a piece of round stock know what a difficult matter it is to do it accurately. This can easily be accomplished, however, by the use of a V-block, and, as these can be purchased for a few cents, the mechanic is urged to procure one. Their use is shown in the photograph, Fig. 45. In the event the mechanic desires to make one for himself, it can easily be done on a shaper, and the sides of the groove are cut at exactly 45 degrees.

The rate of feed and speed for small bench drills should vary with the diameter of the drill and the hardness of the metal being drilled. As a general rule, small drills should be run at high speed and larger ones at lower speed. An easy method of obtaining the approximate speed is that of dividing 80, 110 and 180 by the diameter of the drill, which gives the number of revolutions per minute for steel, cast iron and brass, respectively. In drilling wrought iron or steel, the drill should be flooded with oil or cutting compound (soap and water

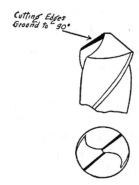

Cutting Edges
Ground to 90°

Fig. 47—How to grind a twist drill for drilling brass or cast iron

make a good substitute). Brass, copper and cast iron should be drilled dry.

When grinding a drill, be careful not to "burn" it by holding it on the wheel too long without dipping it in a convenient receptacle of cold water. "Burning" a drill means excessively heating it until it loses its hardness.

CHAPTER IV

How to make soft solder adhere—Soldering fluxes—Preparation of metallic surfaces to receive solder—Methods of holding work while solder is being applied—Information on silver soldering—Silver soldering outfit—Composition of silver solder—Application of silver solder.

Soldering, both hard and soft, is an important operation with which the model engineer will have to become very familiar. Both processes require extensive practice to become proficient in, but this should not discourage the model maker, as it is quite possible to do good work with a little practice, providing the directions are followed carefully and the necessary precautions to insure success are taken. It may be that the first two or three jobs of silver soldering or brazing will not be entirely successful, but after the model maker has made a few experiments along this line, no difficulty will be experienced in doing good work, which, although it may not be perfect, will serve its purpose.

Before treating the subject of hard soldering, a few words will be devoted to the art of soft soldering. The most important part of soft soldering is that of properly preparing the surfaces to be soldered and holding them rigidly in place while the solder is being applied. The patience of a beginner in soldering is often exhausted when the solder is applied to the surface and repeatedly rolls off without adhering to the metal. Many unkind words are very apt to be said about the various implements employed and the art of soldering in general, under these circumstances, but the workman may rest assured that it is no fault of the solder he is using and, nine times out of ten, it is the method of applying it.

Before solder is applied to a metallic surface, the surface should first be scraped perfectly clean with a small tool that can be easily ground into shape from an old file. Although the surface should be clean and bright, it is not necessary to scrape excessively until a noticeable depression is formed in the metal. The surfaces should be scraped just before the worker is ready

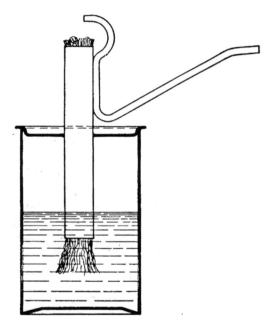

Fig. 48—A simple alcohol torch for soldering

to apply the solder, as long standing will produce a thin film of oxide, to which solder does not readily adhere. Once the surface is cleaned, it should not be touched with the fingers, as this always leaves grease upon the surface no matter how clean the hands are kept. After the scraping is done and the soldering copper is heated, the flux should be applied. A good flux for soft soldering can be prepared by dissolving small pieces of zinc in hydrochloric acid. When this is done, a violent chemical reaction takes place between the zinc and the acid, which results in the formation of a solution of zinc chloride. This is kept in a small glass bottle and applied with a small bristle brush or wooden dauber. In making this solution, the zinc should be added to the acid until no more chemical action takes place.

With the surface prepared according to the foregoing directions, and with the flux in place, the solder is ready

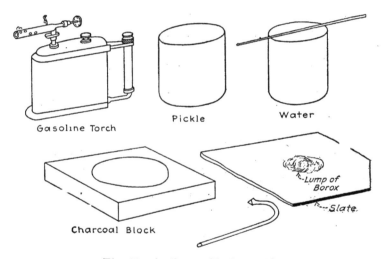

Gasoline Torch   Pickle   Water

Charcoal Block   Lump of Borox   Slate

Fig. 49—A silver soldering outfit

to be applied. Solder in the form of a heavy wire is the most convenient to use, especially for the beginner. The copper should be brought in contact with the work and the solder fed to the tip or point of the copper as fast as it melts and runs. If it melts at the instant it touches the copper, this indicates that the copper is far too hot, and this is a common mistake of many beginners. The copper should be just hot enough for the solder to melt after it has been in contact with it for a short time. After the solder has attached itself to the metal, it may appear very uneven in places, and to remedy this the hot copper is run lightly over the joint to even the depressions and projections. In heating the soldering copper, the tip or end should not be placed directly in contact with the flame, as this burns the tin off and renders it more or less unsuitable for use. The upper part of the copper may be exposed to the flame and the tip will be heated by the thermal conductivity of the metal, which, by the way, is very high.

Although it is quite necessary to employ the ordinary soldering copper in many cases, the best and most effective method is that of applying the heat directly to the surfaces to be joined together. The heat may be supplied by an alcohol lamp, gasoline blow torch or a Bunsen burner. The flame used must be free from soot, otherwise it will contaminate the surface and render it impervious to solder. After the metal is heated in the flame, the solder is applied by bringing it in contact with the heated metal and holding it there until it melts and runs into place. The joint should be given plenty of time to cool before it is handled roughly. Many times it is necessary to bind the pieces together that are to be soldered with iron wire. This holds them rigidly in place until after the solder has thoroughly cooled. In employing the method of direct heating in inaccessible corners, it is best to use a small alcohol burner such as that shown in Fig. 48. This can be made with very little trouble and serves its purpose well. It merely consists of a small metal container with a cotton wick in it, at the end of which the alcohol burns. A small metal tube is soldered to the container so that the end of it comes directly over the wick. By blowing into the tube, the flame can be greatly extended and directed to any part of the work at hand. There is one precaution necessary in soldering by the direct application of heat: The two objects to be soldered together must both be at the same temperature. If a small piece and a large piece of metal are to be soldered together, the small piece is very apt to become heated much more quickly than the larger piece, and the piece that is heated to the greatest temperature will absorb most solder. This should be prevented as far as possible, and can be avoided in many cases by heating the larger piece first.

In soldering certain objects, it is sometimes practical to first wire them together so that they will hold the position that they are to be soldered together in. Of

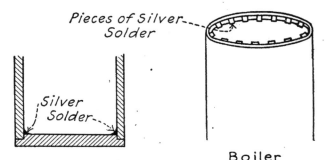

Fig. 50—Right: How the silver solder is laid on the boiler. Left: A finished job

course, the pieces should be so wired that the wire will in no way interfere with the soldering. The wire should not be removed until the work has cooled sufficiently, otherwise the job is very apt to be spoiled.

Many times it is necessary to "tin" a piece of metal before soldering it to another piece, and this operation is very easily done by placing tiny pieces of solder about the surface of the piece and then heating it in a flame until the solder melts. By means of a wire brush or small stick, the molten solder should be spread over the surface. A piece of metal so prepared may very easily be soldered.

Good soldering — and soldering attended with the least possible difficulties — depends largely upon the "flux" or "paste" employed. Many mechanics use their favorite preparation, made according to their own formula, and others prefer the standard market articles, of which there are many that can be recommended. Ordinary resin is best suited for electrical work, owing to the fact that it will not corrode the wire and produces a very dependable connection from the electrical standpoint. Many patented preparations on the market are also

Fig. 51—Proper method of holding a soldering copper and solder

very suitable for electrical soldering. If resin is used, it should be ground up into a very fine powder and sprinkled on the surfaces to be soldered together. Owing to the fact that resin is very soluble in alcohol, a solution of it may be made and applied to the metal in this way by means of a small brush. Immediately this preparation is exposed to the atmosphere, the alcohol evaporates and a very thin film of finely divided resin is deposited upon the surface of the metal.

The process of silver soldering is much more difficult than that of soft soldering, and requires more patience and experience to produce good work. The various tools and materials used in the process of hard soldering or silver soldering are shown in Fig. 49. The outfit, although not elaborate, will enable the model maker to do very good work. It is not the outfit that is so important, but rather its intelligent use. The heat used in silver soldering must be very intense and, for large pieces of work, it is necessary to employ a big flame. The ordinary gasolene blow torch produces a very good flame for this work and it has sufficient heat to melt the solder. The use of the various tools and materials illustrated will now be explained. The acid pickle is made by mixing 1 part of sulphuric acid with 20 parts water. After an object has been silver soldered and cooled sufficiently, it is immersed in this pickle, which thoroughly cleans it and removes all traces of the borax used. This pickle is also used when the work is dirty and greasy, as the solder will not adhere to such a surface, and it is first necessary to clean the metal in this solution. The charcoal block is used to place the work upon while the soldering is being done. The object of this block is to return the heat to the work and this helps greatly in making the operation more rapid. In many cases, however, it is quite impossible to employ this block, even though its use would greatly help the work. The borax is used as a soldering flux just as resin is employed in ordinary soft soldering. The borax is moistened and rubbed on the slate, which produces a paste of borax and water. This is painted on the metal to be soldered at the point where the soldering is to be done. The use of the small scraper shown is obvious. The blowpipe is used on very small work where an alcohol lamp is employed as the source of heat.

Silver solder consists of brazing spelter (brass) and pure metallic silver mixed in varying proportions. The percentage of the metals in the composition determines the melting point, and this may be anywhere from 700°F. to 2000°F. The higher the solder melts, the stronger the joint it produces will be, and *vice versa.* For model boiler work, a solder with a comparatively high melting point should be used. There are other cases where a mixture with a low melting point can be used to advantage. One thing must be kept in mind, however, and this is the necessity of using a solder that is not too close

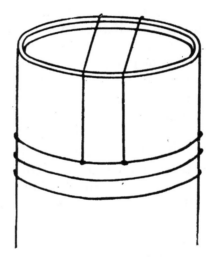

Fig. 52—The end of a boiler wired in place ready for silver soldering

to the melting point of the metals that it is to be used upon. A good solder to use in connection with copper. consists of two parts silver to one part of brass in the form of brazing spelter. A good mixture for work with brass consists of seven parts of silver to two parts of brazing spelter. Silver solder in sheet form can generally be purchased from large jewelers' supply houses. The mechanic can melt up his own ingredients and roll it out into a sheet if he desires. This is the most suitable

form to use it in, as it does not require such a great length
of time to melt.

Assuming that the end of a small boiler is to be silver
soldered into place, the process will be briefly outlined
so the mechanic can obtain an understanding of just
how to proceed. If the metal to be worked upon is very
dirty, it will first be necessary to immerse it in the acid
pickle to completely remove all foreign matter from its
surface. It may also be necessary to scrape the surface
with the smaller tool made for that purpose from an old
file. The end piece is then put in place and held there
by means of iron wire. A little of the borax is then prepared and that portion of the metal which is to receive
the solder on the inside is covered with a thin film of it
by applying it with the brush. Small squares of the
silver solder are then cut with tinner's snips and laid in
places about the bottom of the boiler as near the contacting surfaces as possible. The boiler is then placed
upon the charcoal block and heated. The heat is not
applied directly to that part to be soldered at first, as
this would cause the water in the borax to boil and
would be apt to dislodge the small squares of solder that
were put in place. Instead, the heat is first applied to
the top of the work, and the bottom will become gradually heated by conduction. After the borax has become
sufficiently dried, the flame may be applied directly to
the work and held there until the solder melts and runs
into place. If the end of the boiler was to be soldered
in place from the outside, the solder would be put in
place as shown. If a gasoline torch is used to heat the
work, it should not be brought too close, as the full heat
of the flame will not be utilized if this is done. On the
other hand, if the flame is held too far away, the soot
will be deposited upon the metal and it will then be
necessary to again clean and prepare the surface. After
the solder has melted, the flame should be held on the

work for a few minutes, as this tends to produce a stronger joint. The work is allowed to cool and it is then placed in the pickle and permitted to remain there at least five minutes, after which it is removed and rinsed in clean water. In some cases, the part to be silver soldered may be quite inaccessible, and a small steel rod, with the end split, may be employed to hold the solder if it is in the form of a sheet. The work can then be heated to the proper temperature and the solder held in place by means of the rod until it melts and runs.

A good silver solder for model work on thin brass sheet can be made by mixing twelve parts of silver and one part of brass together. This has a comparatively low melting point and is called "quick" for this reason. Another mixture which is very good for ordinary work consists of six parts silver to one part spelter or brass. This has a much higher melting point than that described previously, but it is much more suitable for some work. In mixing these solders, the mechanic should use care to see that the metals employed are very clean before they are melted together, and it is always safe to clean them with emery cloth before doing this.

CHAPTER V

HARDENING AND TEMPERING STEEL

Simple experiments in the tempering of steel—Proper temperature for tempering to various degrees of hardness—Case hardening—Carbonaceous material employed—Proper heating—Notes on case hardening furnaces.

In model building and experimental work it often becomes necessary to harden a piece of steel and temper it to a definite degree of hardness or soften a hard piece such as a spring so it can be drilled or machined and a few simple experiments in heat treatment of steel are sufficient to enable one to obtain the desired results.

Secure a piece of spring steel wire about $\frac{3}{32}$ inch in diameter and 3 feet long. Heat about 2 feet of it to a dark red color and allow it to cool slowly in the air. This will anneal the wire so it can be hammered easily. Hammer it flat to a thickness of about $\frac{1}{32}$ inch and smooth the surfaces by grinding or filing. Heat the flattened end of the wire to a light yellow color and let the red color extend about 3 inches from the end, allowing a gradual change in color from the end to this point. Cool the wire quickly by dipping it suddenly (endwise) into water as soon as the desired color has appeared.

With a pair of pliers or vise, break off about $\frac{1}{2}$ inch of the end of the hardened piece and notice the grain of the fracture. Break a second piece and compare it with the first, then a third, etc., till the wire bends without breaking, comparing each fracture with the previous ones. You will notice that the part of the metal which was the hottest is always the most brittle and breaks easier than the part which was colder. At some point between these two extremes there is a fracture which is of fine grain and has a silky appearance. That part of the metal which shows the finest silky grain and is too hard to be filed was heated to the proper temperature.

Repeat the operation, this time trying to heat about 5 inches of the piece to the proper temperature, which is somewhere near 1450°F., depending on the quality of the steel. This time test the texture of the grain as before and determine whether the proper temperature was maintained to produce the finest grain and hard metal. When the proper hardening heat has been determined and the

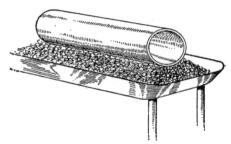

Fig. 53—A small piece of pipe used as an annealing oven

piece hardened properly, tempering is next in order— that is, reducing the hard brittle metal to a tough pliable state suitable for a spring, punch cutting tool, or whatever is desired.

Remove the scale or oxide by grinding or polishing the hardened part of the wire with a piece of emery cloth, soft brick or an emery wheel. Heat slowly about 2 inches of the end and notice the oxide colors as they appear on the surface. As the temperature increases, the first noticeable change will be from the original, polished gray to a light yellow, then a straw, brown, purple, blue, etc., till the piece becomes very soft again.

While heating, move the piece about in the heat so as to draw the end to a gray blue and about 3 or 4 inches from the end to a light yellow. Between these points the entire color scale will appear and indicate to what degree the particular part was last heated. If the piece was hardened properly, the color on the surface indicates the degree of hardness. With a pair of pliers, break and examine the pieces as before. In this case it will be noticed that the part which was heated the hottest is the softest and that the part which has turned to a blue gray will bend easily before breaking. By studying the colors, their corresponding temperatures which are given below and the physical qualities of the steel, any desired degree of hardness may be obtained with comparative ease.

Faint yellow ..	430°F.	Dark brown ...	510°F.
Straw	450°F.	Purple	530°F.
Dark straw ...	470°F.	Blue	560°F.
Brown	490°F.	Gray blue	610°F.

Case hardening or cementation is the process of producing an exterior layer or skin of hardened steel on an article of iron or low carbon steel. This is one of the most useful procedures in metal work, when properly carried out. It enables us to create a hard wearing surface upon material itself incapable of being hardened. We get a comparatively soft and tough interior combined with a hard exterior.

There are several processes, but the same general purpose of impregnating the outside with a high percentage of carbon is obtained. The processes differ in respect to the way it is sought to attain this object. The articles may be heated in an atmosphere of some suitable gas containing carbon until enough carbon has been absorbed by the heated metal. This is the gas process. Then, we may pack the articles in a quantity of ground bone or its equivalent, using a metal box to hold the bone and the work, and then heat the whole to a high temperature until sufficient carbon from the bone has penetrated into the work. This is the method in general use for high-class work, and is recommended to the mechanic.

It is not necessary to have elaborate apparatus. The principal things needed are (1) the packing case of metal, (2) the packing material, and (3) a means of heating the case, when packed, to the proper temperature and hold-

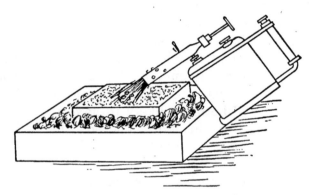

Fig. 54—Simple method of case hardening a small gear wheel, using a blow torch

ing it there for a considerable period—1 to 10 hours or longer, depending upon the result wanted.

After the impregnation with carbon, what we have is a skin or exterior layer of high carbon steel. It remains to harden and, if desired, to temper this skin. If the work is of high grade and there is a desire to have the very best results, then with steel work it will often be necessary to handle the hardening in such a way as to provide for annealing the interior metal. The reason for this is that the high temperature at which the work is impregnated with carbon may have damaged the quality of the steel. The annealing is to restore the quality. It naturally precedes the hardening.

The case or box used may be made of cast soft steel

or of sheet steel. A proper way to make the box of sheet steel is with the aid of the oxy-acetylene welding process. The box may be of almost any size convenient to handle and heat. It must be large enough, since the packing material should be used generously around the articles. For a box 12 x 14 x 24 inches in size, a thickness of sheeting of 0.4 to 0.6 inch is proper. A box such as described should last through perhaps 12 or 15 occasions of its use. What disintegrates the boxes is not so much wear and tear as the action of the atmosphere on the highly heated metal. It may be well to state that it is advisable to make the boxes small rather than large. We must not forget the considerable amount of packing material. A small box will naturally get heated up more quickly than a big one and will be more likely to have the same temperature all the way through. The box should be made with a lid and the two should be so designed that the edges overlap.

In charging a box with the work, it is proper first to coat the inside bottom with a paste made by mixing clay and water. This coat should be allowed to dry thoroughly before going on. After the drying is complete, we put in a layer of the packing material. This layer should be, say, 1½ inches thick. This packing material should be in the condition of a fine powder and should be very dry. The first layer of the articles is put in place. We are careful not to put one against another, but to allow at least 1¼ inches between them. We then pack in the carbonaceous material. Care should be taken to fill in crevices and cavities and other irregularities of the work. We then put on a layer over the work, making this layer, say, 1 inch thick, if more work is to be put in, and 1½ inches thick, if no more work is to go in. In case the box is not full when we have put in 1½ inches of packing material over the top layer of work, then we fill in packing material clear on up to the top. The cover is now put on, fire-clay made pasty with water being used to seal the joints between lid and box.

We are now ready to put the box into the furnace. We put it near the door in order to give the moisture in the fire-clay paste, and any other moisture that may be in the packing material, time to evaporate. After the moisture has evaporated, the box is placed in the hottest part of the furnace.

Almost any kind of a furnace will answer, if it can be made hot enough and provided the temperature in the region where the box is can be managed so that there will be only very small differences at different points. This latter requirement is a very important one. A good gas furnace is, however, easily able to meet it. It is considered undesirable that in the working part of the furnace the temperature at one point should differ by more than 50°F. or 60°F. from that at any other point. If improperly heated, the work may, accordingly, come out of the box in various conditions. Or, if the box is used to hold one large article, then one part of the article may have received more carbon than another.

A proper temperature to use in case-hardening is 1740°F. This is a *light orange* color verging to a yellow. Case-hardening may be done at a lower temperature—as low, it seems, as a *light red,* under favorable circumstances. It should be remembered that it is not simply the box that has to acquire the temperature but the work itself; and that a sufficient time must be allowed to get the depth of impregnation desired.

One of the simplest materials is that made according to the following formula:

Wood-charcoal .... 9 parts
Common salt .... 1 part

In using this mixture, it may be necessary to go to high temperatures to get results in a reasonable time. Temperatures up to *light yellow* may be used. Another mixture, claimed in authoritative quarters as better, is the following:

> Powdered wood-charcoal .... 6 parts
> Barium carbonate .... 4 parts

With this mixture, a mild tool steel coating, very thin, is obtainable at an *orange* heat or higher; and a high-carbon tool steel coating, $\frac{1}{32}$ inch or thicker, at high temperatures near *light yellow*. An advantage of this mixture is that it may have its activity restored after it has suffered from use, the simple means of restoration being exposure of the material in a thin layer to the influence of the air.

There are many substances used for packing material —for example, wood-charcoal, leather, bone, common salt, sodium carbonate, saltpetre, resin, sawdust, soot, etc. These are used in various combinations.

Case hardening to a very small depth may be accomplished by putting the cold, or moderately heated, article into a bath of potassium cyanide. The bath should be heated, say, to a *bright cherry red* prior to the immersion of the work. The article may be hung by a fine iron wire and allowed to remain until it acquires the temperature of the bath. This will, in some ordinary cases, require about a quarter of an hour or a little longer. It is necessary to point out that cyanide of potassium is a very deadly poison and that even the fumes from it are poisonous. The cast iron pot or other crucible holding it should be enclosed with a hood connecting with a chimney or ventilating shaft.

CHAPTER VI

Abrasive equipment for the model engineer's workshop—Grinding and polishing—Grinding attachments for small grinding head—Bonds used in making abrasive wheels—How to choose a wheel for certain work—Precautions to be taken in mounting wheels.

There are many instances, in certain work, where the short-cut lies in grinding, and owing to the scarcity of published data on this very important subject many remain ignorant of the great utility of simple abrasive materials and equipment in their application to mechanics.

Every workshop should contain a small grinding and polishing head. The one shown in Fig. 55 is a very good machine for the work generally required in the small shop, as it can be used for grinding, polishing and buffing. The machine should be belted to a ⅙-H.P. motor of sufficient speed. In the event the mechanic is unable to procure such a motor and has a small bench grinder, the arrangement shown in Fig. 60 may prove to be of interest to him. The clamping disc of the bench grinder is replaced by a wooden pulley with a groove large enough to accommodate a sewing machine belt. The polishing head is placed close enough to the bench grinder so that one may work conveniently at the former while driving it by means of the bench grinder. If a motor is used, a rheostat would make a very valuable addition to the outfit, as a variation of speed is desirable for different classes of work.

The grinding head should be provided with an assortment of wheels of various shapes, sizes and grits, as every wheel should be adapted to the particular kind of work it is to be used for. A wheel 3 inches in diameter by ½ inch thick is a very good size for general work when used with the small polishing head shown in the picture. As such wheels can be purchased for about 40 cents each, it is advisable to have four or five on hand of various

Fig. 55—A grinding outfit for the model engineer's workshop

grits; from very fine grit to coarse grit. Several round edge wheels of varying thickness should also be on hand, as there are many different jobs and operations where such wheels can be employed with great convenience, as in the cutting of grooves, etc. For very fine and accurate work, small wheels of fine grit should be employed. Such wheels are commonly known as jewelers' wheels, and owing to the difficulty of procuring them with an arbor large enough for use on a half-inch spindle, the little "kink" shown in Fig. 56 may be used. The wheel is clamped between two washers by means of a 10-24 machine screw and nut. The protruding end of the machine screw is then placed in the chuck of the polishing head. As these wheels can be purchased for 10 cents apiece, it is advisable to have an assortment on hand.

Owing to the inability of the small grinding head to stand up under heavy work, a spindle equipped with a larger wheel should also be included in the abrasive equipment of the shop. A small bench grinder is capable of accomplishing quite heavy work and, in many cases, is sufficient for the small shop. The design and construction of a small, heavy-duty grinding head also pre-

Fig. 56—How a jeweler's wheel can be used in the chuck of a small grinding head

sents a nice job for the young mechanic who is desirous of equipping his shop with a minimum of expense.

Polishing, as well as many other operations, can be done two ways: good and bad. It is very easy to do bad polishing and not very difficult to do good polishing providing the proper precautions are taken and the most practical methods used.

Two felt wheels 3 inches in diameter by 1¼ inches wide should be obtained from a polisher's supply house. These can be attached to the tapered end of the polishing head, care being taken that they run true. The periphery

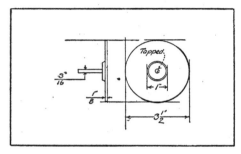

Fig. 57—Drawing of a grinding disc for use with a small grinding head

of each wheel is then sized with a thin coating of hot carpenters' glue and rolled in carborundum or emery powder. The wheels are then put away to dry. One wheel should be prepared with a very fine abrasive powder and

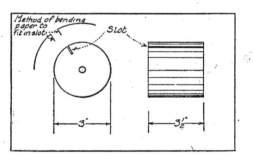

Fig. 58—A cylindrical grinding attachment made from wood—The abrasive paper or cloth is held to its surface as shown

the other with a more coarse powder. When it is necessary, the wheel with the coarse powder can be used to produce a preliminary polish and the other wheel can be used to put the final polish on the work. If carborundum grains are used, the work should be held very lightly against the wheel, as these particles are extremely sharp and cut very easily. Emery is less abrasive in its nature, and it is necessary to bear more heavily on the wheel.

An accessory for use in polishing small flat surfaces is shown in Fig. 57. This is a brass disc 3½ inches in diameter and ⅛ inch thick, provided with a ³⁄₁₆-inch stud or shaft in the center, which is held in the chuck on the spindle of the polishing head. The abrasive paper or cloth discs used on the surface may be cut from the standard sheets obtainable at any hardware store. The disc is fixed to the surface of the brass by means of hot beeswax. This is a very handy little contrivance, especially in polishing small instrument parts.

Another accessory easy to construct and very useful is shown in the drawing. This is a small wooden form or wheel turned out on a lathe with the dimensions as shown. A slot is cut across its face with a hack saw to a depth of ½ inch. A piece of abrasive cloth of any desired grit is cut to a length just exceeding the circumference of the form and the overlapping ends are bent at right angles and forced down in the slot. This holds the cloth to the surface of the wheel.

A very convenient little contrivance is shown in Fig. 59. This can be used for polishing small flat surfaces quickly. As will be seen from the photograph, it is merely a board equipped with a clamp at each end. The ends of the abrasive paper or cloth are placed under the clamps and held tightly by screwing down the winged nuts. To polish small, flat objects it is only necessary to lay them on the abrasive cloth and rub them briskly over its surface with an oscillating motion. Aside from a quick and convenient means of producing a polish, it is also useful in getting work down to exact size, as rapidly revolving abrasive surfaces generally cut too fast for extreme accuracy by hand. The board is made wide enough to accommodate two strips of abrasive cloth each 4½ inches wide by 11 inches long, which is just half of a standard 9 x 11 sheet.' One strip should be of a very fine grit and the other should be of a coarse grit.

It is advisable to have both a coarse and a fine grit stone in the shop. In place of these, the writer would

Fig. 59—A lap board which is very convenient for polishing and grinding . flat surfaces

recommend the use of a combination stone which is composed of two stones (coarse and fine) cemented together. Such a stone is both convenient and economical, as it can be purchased at the price of a single plain stone and thereby saves the expense of an extra one. The tool is first edged on the coarse side and the fine side is then used to further remove the imperfections of the previous operation. Water or oil may be used to lubricate the stone. Some mechanics prefer one and some the other. It is optional which is used. If it is desired to produce an especially keen and delicate edge, the surface of the stone should be tempered with wax or vaseline. These substances fill the pores and interstices of the surface and regulate the sharpening process.

A small hand stone should also be in every mechanic's tool kit. This will be found very useful in sharpening small drills, reamer edges, compass points, etc.

The following words on the technique of the grinding wheel should prove to be of interest to the average mechanic, as the information given will help him to choose "the right wheel for the right place," as one big manufacturer puts it.

Grinding wheels should be adapted to the particular kind of work they are to be used for. Shape, grade, grit

Fig. 60—Driving a grinding head with a hand grinder

and bond should be considered when choosing a wheel for a certain class of work. The importance of this is obvious to the careful mechanic.

Grit—The grit of a wheel is determined by the size of the abrasive particles that compose it. Coarse grit wheels are composed of large particles and the finer wheels of smaller particles. The grit is determined by the number of meshes to the linear inch of a sieve that the particles composing the wheel will pass through. These numbers generally run from 12 to 150. Low numbers indicate the coarse wheels, and high numbers the finer grit wheels. Where large, heavy cuts are to be made, wheels of a coarse grit should be used. In more delicate operations where great accuracy is sought, wheels of a finer grit should be used.

Bond—The bond of a wheel is the substance used to hold the abrasive particles together. The nature of the bond as well as the regulation of the mixing and baking process determines the degree of hardness of a wheel. The bond of a wheel is *very important* and should always be considered. The degree of hardness or the bond of a wheel determines how rapidly the particles composing it will break away from their settings. If they break away from their settings rapidly, the diameter of the wheel is reduced correspondingly fast, and the wheel is said to be "soft." Such a wheel will cut fast and freely if its surface velocity is sufficiently rapid, as new and sharp abrasive particles are continually exposed owing to the old ones breaking off easily. There are many classes of work where such a wheel is necessary, and there are also many operations where it could not be used at all. If a wheel is "hard," its particles do not leave the bond so easily, and therefore its diameter will not be reduced so rapidly. However, if such a wheel is run too slowly and made to take a heavy cut, it will "glaze" badly. This is caused by the abrasive particles losing their cutting power by wear before they are able to break away from their setting owing to the hardness of the bond. There are cases where the employment of such a wheel is necessary. As an example: A 3-foot steel shaft is being turned to an exact diameter between centers on a universal grinding machine and a cut of $\frac{1}{100}$th of an inch is being made the entire length of the shaft. If the wheel used is not hard enough, it will reduce in diameter in making the cut and inaccuracy will result, as one end of the shaft will be larger than the other.

A list of the common bonds used in abrasive wheels follows, together with the special class of work each bond is most adaptable to.

Vitrified—The bond of such a wheel is generally a special clay. Vitrified wheels are used for cylindrical

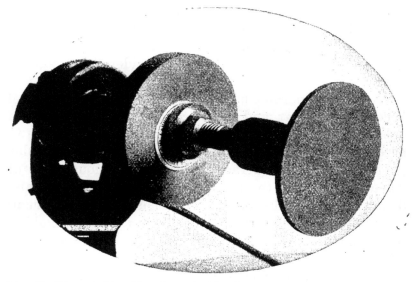

Fig. 61—The grinding disc shown in Fig. 57 attached to a grinding head

and cutter grinding as well as for general machine shop work.

Silicate—The bond is silicate of soda, and such wheels find wide use in all shop work.

Elastic—The bond is pure shellac, and as it produces a wheel that is not brittle, it is generally used in making very thin wheels (as thin as $\frac{1}{16}$ inch). Such wheels are used for saw gumming, grinding reamers, cutters and arbors. They are also used in cutting off small stock such as steel tubing, etc.

Rubber Wheels—The bond consists of rubber and sulphur. Although rubber wheels do not find wide use, there is special work where they are required, as they are not brittle and may be run at high surface velocity. Rubber wheels are also capable of withstanding great lateral pressure.

It must be understood that the wheels of each bond are made in several different grades in varying degrees of hardness. The amount of metal to be removed, the physical nature of the metal being worked upon, and the desired condition of the finished surfaces are the three governing factors that should be considered in choosing a wheel.

After considerable use, the profile of a grinding wheel becomes irregular and is restored by a process called dressing. This is very important, as it is impossible to accomplish accurate work on a wheel with an irregular profile or grinding surface. There are various types of wheel dressers on the market, each with its merits and disadvantages. They generally consist of a set of small steel wheels which revolve rapidly in a suitable holder, when they are placed against the face of the grinding wheel to be dressed. Such wheel dressers are very well for large grinding wheels, but they will be found unsuitable for use on small wheels such as used in the home shop. For dressing such wheels, a piece of an old broken wheel can be held to the edge of the wheel to be dressed, and if a little care is used very good results can be obtained in bringing a true surface to the wheel. This process is also a good remedy for a "glazed" wheel.

A still more satisfactory stunt is to revolve the grinder at low speed and true its periphery with a flat file held end on to actually "turn" the surface.

Warning—The following words are intended for those

mechanics, who have large, power-driven grinder heads in their workshops:

Grinding wheels sometimes burst without any warning and many men have been instantly killed by being struck with a flying piece. This is generally caused by carelessness, and with a little precaution many serious accidents could be prevented. Wheels should fit freely on the spindle and should never be forced on. The reason for this is as follows: If a lead-bushed wheel fits too snugly on the spindle and the bearing of the grinding head becomes too hot, the heat will be communicated to the lead bushing, and as the lead has quite a high coefficient of cubical expansion it will bring a considerable pressure to bear upon the arbor of the wheel and in many cases burst it. The wheel should fit the shaft or spindle with sufficient freedom to permit this possible expansion of the lead bushing. Wheels should also be sounded carefully before being mounted and the flanges should never pinch the wheel too tightly. Wheels should never be run above the rated surface velocity.

CHAPTER VII

PATTERN MAKING

General foundry practice—How moulds are made—Various kinds of patterns—Making patterns—Cores and core boxes—Parted patterns and how to make them—Finishing patterns.

*P*attern making is an extensive trade, and a man could well spend a lifetime learning its various sides; the beginner, therefore, should not attempt the building of large or complicated pieces without the help and advice of a practical man, but by keeping constantly in mind the elementary operations in the moulding and drawing of ordinary patterns he should be able to turn out satisfactory work, and not suffer the humiliation of hearing it pronounced faulty by the foundrymen.

In this treatise no attempt will be made to describe to any length the tools of the trade and the mode of using them. Ordinary carpenters' tools will answer the purpose for the amateur, and it is assumed that he is reasonably familiar with their use. Many patterns do not require the use of a lathe in making them, but for those that do, even though not expert in the use of turning tools, the operator will usually be able to "scrape" to shape and sandpaper his work so that it appears presentable. Screws, nails and brads of various lengths and sizes should be at hand, and glue and shellac must be provided. Generally but a small quantity of glue is required, in which case the ready-made liquid glue will prove more satisfactory than the solid kind, which has to be heated in water and melted. Several sizes of wooden screw clamps would be helpful in holding the parts together after gluing. Shellac is used for the finishing of pat-

terns. Either the white or orange may be bought prepared. Several coats are generally required, as the preliminary ones raise the grain of the wood and roughen the
surface, requiring the application of fine sandpaper after
each coat; the finished surface must be smooth, in order
that the pattern may be easily "drawn" from the mould.
Here a caution should be inserted regarding the use of
sandpaper; the article must be brought as nearly as possible to its final form and finish without the use of this

Fig. 62A   Fig. 62B

Fig. 62A—A pattern on a moulding board
Fig. 62B—A two-piece pattern

abrasive, and great care should be taken not to curve
supposedly flat surfaces or to round supposedly square
corners through employing an excess of zeal in its application.

Almost any fairly soft, evenly grained wood capable
of taking a good finish, and which will not warp or swell
to an undue extent through coming in contact with damp
sand in the mould may be used for this work. Cherry
and mahogany, because of their freedom from warping,

shrinking and swelling, and because of the fine finish they are capable of taking, are considered about the best for regular commercial patterns, but they are rather expensive, and cheaper lumber, such as white pine or redwood, will answer the purposes of the amateur just as well.

In order to go about this work intelligently, it is essential to understand clearly the series of operations by

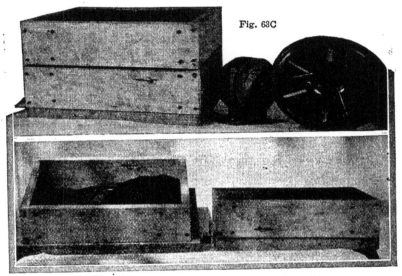

Fig. 63C

Fig. 63A

Fig. 63B

Fig. 63A—(Lower). A pattern partly covered with sand
Fig. 63B—(Lower). Drag completely filled and smoothed
Fig. 63C—(Above). Cope in position filled with sand

which a casting is produced at the foundry, and the reader's attention is called to the series of photographs representing the process of making a sand mould from a simple one-piece pattern, that of a face plate to be used on a small lathe or drill press. Small castings are moulded in a "flask," which in its plainest form consists of two rectangular frames resting upon a loose piece called the bottom board. The upper frame is called the "cope," and the lower is known as the "drag." A "moulding board," which for our purpose may be considered a duplicate of the "bottom board," is also provided. In the photographs the "flask" is represented by two wooden frames resting upon a bottom board. A regular flask, however, would be larger than this and of more complicated construction, being provided with guide pins between cope and drag, so that they will always fit together properly, cope "bars," handles, etc., and they are often built of iron. For the purpose of simplification all such details are omitted from the pictures. Sections

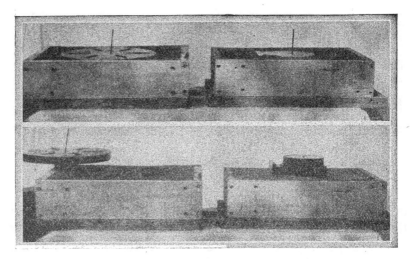

Fig. 64—(Above). Pattern exposed and piece moulded in place
Fig. 65—(Below). Pattern withdrawn and two-piece pattern held together with dowel pins

called "cheek pieces" are sometimes introduced between cope and drag for producing complicated castings and a number of pieces are usually moulded in one flask, which is partitioned off for the purpose. While the procedure that follows may not always be adhered to in regular foundry practice, the beginner by so constructing his patterns that they can be moulded by such a series of operations is assured of good results.

Our face plate pattern will first be placed upon the "moulding board," flat side down, as in Fig. 62A, then the drag, shown, on end, behind the moulding board, is placed over it as in Fig. 63A, and filled with moulding sand which is "rammed" down and smoothed off even with top of drag. Fig. 63A shows the pattern partly covered with sand, and Fig. 63B represents the drag completely filled and smoothed off, the pattern being at the bottom out of sight. Next, the bottom board

Fig. 66A   Fig. 66B

Fig. 66A—Pattern in place on board
Fig. 66B—Typical parted pattern with horizontal core print

is placed upon the top of the drag for the moment, and the whole is turned over, the removal of the moulding board exposing to view the side of the pattern which rested upon it, as shown in Fig. 64. Now the cope is placed upon the drag, the pattern still being in place, filled with sand and rammed, and lifted off again. A "sprue pin," whose purpose is to form an opening through which the molten metal may be poured into the mould, is set into the cope previously to filling in the sand, but this has been omitted from the pictures. Fig. 63C shows the cope in position, filled with sand.

The cope having been filled and lifted off, the pattern is removed by driving in a "draw pin" and "rapping" to loosen it from the sand. Sometimes considerable rapping is necessary, and it is important to remember that *a pattern should be as strongly constructed as possible in order to withstand this treatment.*

Fig. 65 shows the appearance of the mould after withdrawing the pattern. A "gate" or channel is cut through the sand from under the sprue pin opening to the mould, vents for the escape of air are provided, and then the cope is replaced and all is ready for pouring. Assuming that the reader has these fundamental operations

Fig. 67A   Fig. 67B

Fig. 67A—Pattern drawn from sand
Fig. 67B—Parted pattern with core print

clearly in mind, we will now take up the question of "draft."

"Draft" refers to the tapering of the sides of a pattern so that it may be easily "drawn" from the mould without breaking out the sand. In Fig. 70A is shown a section of a plain flange pattern in position upon the moulding board ready to be covered with sand. The sides of the pattern are tapered off in order to facilitate its being "drawn," as shown in Fig. 70B. It will be seen that if the sides were tapered in the opposite directions the pattern could not be withdrawn without pulling the sand up with it and spoiling the mould, and that if the sides were made perpendicular difficulty would also be experienced. Note that this pattern is made with a hole in the center, and that the sides of the hole are so tapered that upon drawing, a column of sand will be left

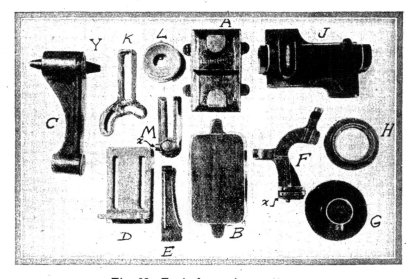

Fig. 68—Typical one-piece patterns

standing in the center of the mould, so that the molten metal running around it will form the hole in the casting. Such a column of sand formed directly by the pattern is known as a "green sand core." The subject of cores will be taken up later.

Every side of a pattern which in the original design or drawing is shown as perpendicular to the plane upon which the pattern will be moulded, or, in other words, every side which must slide through the sand in drawing from the mould, *must be tapered or given draft, and this taper must be in the right direction.* In Fig. 70A the amount of draft is purposely exaggerated; one-eighth of an inch to the foot is usually allowed for small articles, but the amateur could well provide a greater amount, at least enough to make the taper easily discernible to the eye, for this would mean less work for the moulder and better castings. He should always keep in mind the idea of the moulding board, as illustrated in Fig. 62A, and the drawing of the pattern as shown in Fig. 66, and before beginning to make a pattern of any kind should always ask himself: "How will it be moulded? How will it be drawn? Are there any parts which will not draw without ruining the mould?" Upon deciding the direction in which it will be drawn, he will then be able to determine which sides should be given draft and its cor-

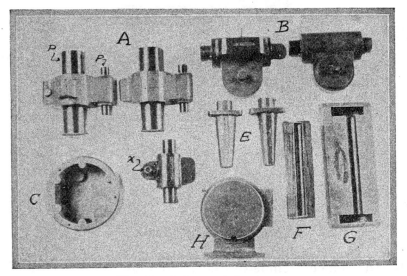

Fig. 69—A few parted patterns

rect inclination. Imagine the pattern to be placed on the moulding board as in Fig. 62; then the direction of drawing, in the present position, would be downward; therefore, all outer sides of the pattern should slope *upward* and *inward* and the inner sides, as the sides of the hole in flange in Fig. 70B, should incline *upward* and *out-*

ward. This, of course, includes curved as well as straight surfaces.

Keeping these facts in mind, there should be no difficulty encountered in constructing one-piece patterns having one side flat to allow for placing upon the moulding board. It is not necessary, however, that the flat side be made smooth, that is, having no openings or recesses. Fig. 71 represents a pulley pattern set upon the moulding

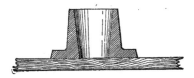

Fig. 70A—Flanged pattern showing draft

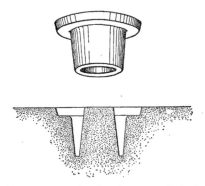

Fig. 70B—Pattern removed, showing core left in sand mould

board. The bottom is recessed and a lower hub is formed whose end is higher than the pulley rim in the position shown. This will cause no trouble, provided draft is given the hub and inside rim as indicated; the drag will be filled, rammed and turned over as before, then the cope will be set in place and filled and some of the sand will fill up this recess, now on the upper side, but will be lifted off with the remainder of the cope sand. The pattern is then removed, and upon replacing the cope this projecting part of the sand will extend down into the mould in the drag, thus forming a recessed casting of the same shape as the pattern. Many other patterns fall under this class, such as the base plate of Fig. 72, with its recessed under side and interior bosses and webs. This pattern is also shown at B, in Fig. 68, and beside it is shown a bottom view of a similar one (A, Fig. 68).

Care should always be taken not to build a pattern which cannot be drawn, such as the pulley of Fig. 73, which is identical with the one of Fig. 67, except that the bottom hub projects beyond the pulley rim, and if moulded as shown the sand between rim and surface of drag would be torn out when drawing the pattern. It is not impossible to produce a casting from a one-piece pattern like this one, but it would not be wise for the beginner to make one of this type without obtaining practical advice upon the subject. The safest and best way would be to make that part of the hub which projects below the rim and interferes with the use of the moulding board, removable, and held in place by dowel pins. This would really form what is known as a "parted" pattern. *All hubs, bosses or other projections upon the side of a one-piece pattern which would interfere with the use of the moulding board as shown in Fig. 62, should be made removable.*

A group of typical one-piece patterns is shown in the photograph, Fig. 68. A is the under side of a base plate similar to the one of Fig. 72, a top view of which can be seen at B. C is a drill press arm; note that the boss Y and boss with tapered "core print," Z, are removable, for the pattern would be placed on the moulding board with that side down. (Core prints taken up later.) D is a slotted angle plate; the slots must be given draft and the piece at right angles to slotted base should be given considerable draft, for this part is down in the drag. Note rounded corner piece, or "fillet." At E is a crank handle; F a follow rest for a lathe, and would be placed on the board laid flat as shown. X is a "loose piece," to be described later. G and H are flanges; K should be specially noted as bearing out what was said in the introduction; it was made to replace the broken part of an old lathe for which a new casting could not be obtained. The

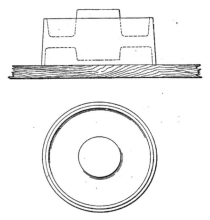

Fig. 71—A pulley pattern

pattern was cut out by hand in about an hour, and the casting obtained only required a little hand filing and cost about twenty cents. At M is the tool rest slide for a speed lathe, X being another "loose piece." These patterns are all cast like the face plate shown in Figs. 62, 63, 64 and 65. The group gives an idea of the great variety of forms under this class.

But many patterns have to be made in halves, or in two or more parts, held together by wood or brass dowel pins, called "pattern pins." Such patterns are known as "parted patterns," and are so made that one part may be moulded in the drag and another in the cope; the dividing or "parting line" of the pattern *coincides with parting line between cope and drag*. If made and moulded in one piece such a pattern could not be removed from the mould without completely breaking up the sand, but the two parts are so constructed that each may be individually drawn, and then when the cope and drag are again fitted to each other, the mould formed by the impressions of the two pieces will be of the exact shape of the pattern.

Photographs Fig. 62B and Fig. 63C (center) show a pattern which, on account of the shape of its base, obviously could not be cast in one piece; it is therefore "split" through the middle and moulded as illustrated in the series of photographs 62B, 63B, 64, 65. One part, that without dowel pins, is laid upon the moulding board, as in Fig. 62B; the other part with pins may be seen

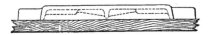

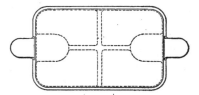

Fig. 72—A base plate pattern

standing on end. The procedure is the same as the moulding of the one-piece face plate previously described, until the drag is turned over and moulding board removed, exposing the flat side of the pattern to view. Now at this point the second part of the pattern is placed upon the piece still in the drag, the pattern pins holding it in the proper position, see Fig. 65. Then the cope is set in place, filled with sand, as in Fig. 63C, and removed. As the pattern pins fit loosely, the upper part of the pattern

comes off with the cope which is turned over, and this
part and that in the drag removed by means of "draw
pins." Fig. 64 shows the piece first moulded in the drag
ready for drawing, with draw pin in place. The cope is
now replaced upon the drag after providing gate, vents,
etc. (not shown) and the mould is ready for pouring.
The important point to remember then, in making a
parted pattern, is to provide the flat side of one piece
with dowel pin holes, but no pins, in order that it may
be laid flat upon the moulding board exactly as though

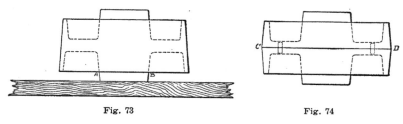

Fig. 73   Fig. 74

Fig. 73—A pattern for a pulley that cannot be drawn
Fig. 74—A parted pattern for the pulley shown in Fig. 73

it were a one-piece pattern; therefore *the same rules for
draft should be applied.* Then supposing that both parts
were laid flat, side by side, *the taper or draft given the
sides of the two respective parts should be in the same*
direction. A study of Figs. 62B, 63B, 64, 65 will make
this clear, and Figs. 62A and 62B will show how the
moulding board is used for the moulding of parted as well
as one-piece patterns. By keeping in mind the use of
this board one should be enabled to reason out the correct
way to go about the building of patterns of either type.

The pulley illustrated in Fig. 73, when provided with
a loose end on the lower hub, is a regular parted pattern;
but now that this class has been explained a better way
to make it will be described; by having the parting line
of the pattern, and therefore of the mould, directly

through the center of the rim as in Fig. 74, much less draft need be given the rim, since now only half of its width has to be drawn through the sand; this also means less work for the moulder, for the pattern will draw easier and not require so much rapping; a better casting

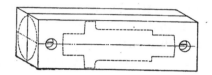

Fig. 75—A parted pattern ready for turning

results, and there will be less machine work and waste of material due to its more symmetrical shape.

A few parted patterns are shown in Fig. 69. For the moment no attention should be paid to the black projections seen on some of them; these are "core prints" and will be explained later. At A, B, and E, are three pairs, and at D half of another. The cylindrical parts of these are lathe turned, and the other pieces fitted and glued.

Fig. 76—A steel "spur center" which is used as a live center in wood turning

E could be made in one piece and cast vertically, for the "nose" has considerable taper, were it not for the cylindrical projection above it (this one is not a core print) which would have to be pinned on so that it could be removed and allow the "nose" to be stood upon the moulding board resting upon its flange. H is a half of the pattern shown as being moulded in Fig. 62B. C, F and G are "core boxes," and will be taken up further on.

It must not be supposed that parted patterns are built in one piece and then sawed in two afterward; Fig. 75 shows two pieces planed and fitted together with dowel pins, the regular pins that remain in the pattern, and temporarily secured by counter-sunk screws at the ends. Enough material should be allowed at the ends for the pattern to be turned to shape and finished at the ends without running into these screws; the assembled piece must be wide enough to allow of its being turned to correct diameter. Great care is necessary in placing the work between centers, for the centers should fit directly into the parting line; otherwise the halves will be found unsymmetrical. At Fig. 76 is a steel "spur center" used as a live center in wood turning; the spur holds the end of the work firmly and no dog or other holder is required. *The dowel pins must always be fitted first, before turning to shape,* for it would be most difficult to put them in accurately afterward. They should fit loosely, so that the parts will fall apart of their own weight, but not so loose that the pieces can shift sideways.

Sometimes it is necessary to make a pattern in three or more parts; on other occasions an irregular parting surface must be made in the mould; often the pattern parting line does not coincide with that of the sand, but these more complicated operations would be beyond the scope of this chapter, and the beginner is advised to stick to the easier forms for a while. Meanwhile, the subject of "core prints" and "core boxes" requires some attention.

Many patterns, while varnished or finished "bright," are seen to have cylindrical or other projections painted black, or if the body of the piece is black, the projections are left "bright." In every instance these strange looking projections are found to occur at points where holes or openings or recesses are required in the casting, and one is inclined to wonder why holes to correspond were not made directly in the pattern itself. In some cases this could be done, as, for instance, in the.patterns shown at D, H, K and L of Fig. 68, and in the case of the slots of the face plate in Fig. 62A. It will be noted, however, that all these holes and slots are shallow; if they were longer and of small diameter the draft would have to be

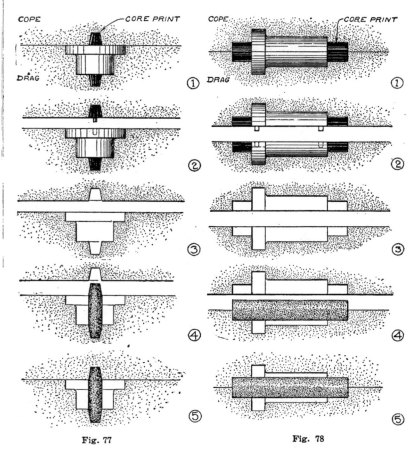

Fig. 77

Fig. 78

Fig. 77—Moulding pattern with vertical core prints and placing of vertical dry sand core

Fig. 78—Moulding of parted pattern with horizontal cylindrical core prints, horizontal core

excessive to allow the pattern to be drawn at all, and in a great many cases it would be impossible to use a "green sand core" due either to the length of the hole or to the fact of its occurrence below the parting line of pattern or mould, as in Fig. 86. In these instances "dry sand" or moulded and baked cores must be used, and these make necessary the employment of "core prints."

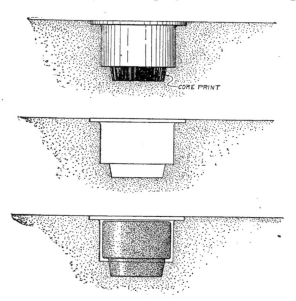

CORE PRINT

Figs. 79, 80, 81—Moulding the pattern of a steam gauge or meter case with opening on one side only

"Core prints" are those parts of a pattern which form depressions in the sand mould, these depressions being used to support a "core" made of sand or sand and glue, moulded in a wooden "core box" and baked hard in a "core oven." When the metal is poured into the mould, it flows around this core and a hole is thus formed in the casting. The remains of the core are removed with the casting and dug out of the hole. Core prints do not necessarily have to be of the ·same size as the hole to be

formed in the casting; they may be of any shape that will draw, but the body of the core, allowing a small amount for shrinkage of the metal, *must be the exact size and shape of the hole or other opening required in the casting.* If the hole is to be cylindrical, the core body must be cylindrical; if a square or rectangular core is used, no matter what the shape of the core prints, the casting will

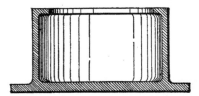

Fig. 82—Cross sectional view of finished casting. This should be compared with the core shown in Fig. 81

be made with a square or rectangular hole, and so on. Then end or ends of the core must fit closely into the core print depressions in the mould, otherwise the molten metal might run between and cause the core to shift in the mould. Now for an example:

The face plate as shown in Fig. 62 is being cast without a hole in the center. Suppose that in order to make the machine work easier it is decided to have a hole "cored" in the casting, but that the hole is too small in diameter in proportion to its length, to make feasible the use of a "green sand" core like that of Fig. 70A. Since the hole is too small in diameter in proportion to its length to do this, what is known as a "vertical" core will be used. If the core prints were cylindrical, they would be hard to draw, therefore they are tapered or given draft. Fig. 66A shows the pattern in place on the moulding board with the top core print, which is merely a piece of wood shaped like the frustum of a cone and painted black, fastened in place. If the bottom one was similarly fastened, it would interfere with the use of the moulding board, so it is made removable. This piece is shown standing in front of the pattern. Its bottom has a pin which fits into a hole in the bottom of the main pattern. No set rule is used in tapering these vertical core prints; one often used for small ones is to make the length equal to the diameter of the core, large end of same diameter and the small end half the diameter.

The core itself is moulded in a "core box," which sometimes is made in halves pinned together like a parted

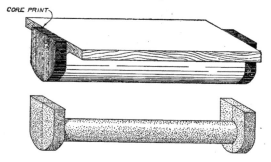

CORE PRINT

Figs. 83, 84—Pattern and core prints. Pattern will be moulded with flat (top) surface resting on moulding board. Appearance of core made in core box.

pattern. Such a core box is seen standing at the left of the pattern in the picture. When the core is to be symmetrical, as this one is, it is really only necessary to make half of the box; then at the foundry they will mould two half cores, bake them and "paste" them together, thus forming a whole core and saving the pattern maker considerable work. A half-box must be accurately made in order that the two half cores will fit together. As a matter of fact, it is hardly necessary to make a core box at all for either a vertical or horizontal *cylindrical core,* unless it is of an odd size, for all foundries keep on hand what are known as "standard" cylindrical cores of all diameters, from a quarter-inch up, running by eighths or sixteenths. These are cut to the required length, and if a vertical core with tapered ends is required, they "rasp" off the end to the right taper. There is no extra charge for the use of these standard cores, for the making and baking of special cores is not then necessary, and the pattern maker is thus saved considerable time and labor, especially if he is not provided with a "core box plane," and has to shape his core boxes with gouge and sandpaper.

To return to the face plate: This is moulded in precisely the same way as before, the loose-bottom core print being handled the same way as the corresponding part of a parted pattern. Before the mould is closed up for pouring, the baked core is set in place in the core print depression formed in the drag by the upper print shown in Fig. 66A; then when the cope is put on, the other core print depression fits down over the top of the core, holding it firmly in place. This procedure is shown graphically in Fig. 77, which represents the successive operations of moulding a flange with a vertical core.

Fig. 78 shows the moulding of a longer flange pattern, which is best made parted and moulded horizontally, thus requiring a horizontal core. In this case, the core prints would be made cylindrical, the same diameter as the core, and, of course, parted like the rest of the pattern; in making a pattern of this type, the core prints would be turned up as an integral part of the pattern. A good rule for small core prints of this kind is to make the length equal to the diameter. The core is made similar to the vertical one, either in one piece or moulded in a half core box and the halves pasted together. In Fig. 69, F, is a half box for a cylindrical horizontal core. Although such a core box is easier to make than the kind shown in Fig. 66A, the amateur is advised to allow the use of the standard cores at the foundry whenever possible.

Photographs 66A and 67B illustrate the moulding of a typical parted pattern with horizontal cylindrical core prints. The complete pattern is shown in Fig. 66B, and to the right of it is a casting which was made from it. Note the hole through the casting. Standing up_ right behind the pattern is the core. In Fig. 67A the

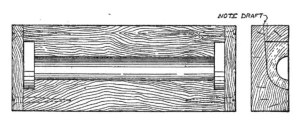

NOTE DRAFT

Fig. 85—Plan and end view of half core box

cope has been lifted, the half pattern with it, turned over and the pattern drawn. Note that the core prints on this piece have left two semi-cylindrical depressions in the sand. At 67B the other half of the pattern has been removed also, leaving a half mould in the drag and two semi-cylindrical core print depressions into which the cylindrical baked core has been placed. This done, the cope will be replaced and the mould will be ready for pouring. The casting shown upon the corner of the drag in Fig. 67B is not, of course, supposed to have been taken from the mould; it was placed there merely to give an_ other view of the one of Fig. 66B.

Sometimes a number of holes, vertical, horizontal, or inclined, or recesses and openings of various shapes are cored into the same casting; core boxes are often ex_ tremely complicated, and, in fact, castings are frequently made without a pattern, the mould being made up of cores alone. If the casting is to be open at one side only, as in the case of the box-like steam gauge or meter case shown in Fig. 82, but one core print is used, and the vertical core rests upright in the depression formed by it, the core, of course, being a "dry sand" one made in a special core box. The core in the picture is short and thick, and is easily supported in this way, but if it had to be placed horizontally or if it had been long and thin, the inner end would have had to have been held in place by one or more short inner cores and the holes formed in the casting plugged up later. The water jacketed cylinder of a gas engine would be illustrative of such a case. Space will not permit the discussion of such complicated cores and core boxes, which would prove troublesome for the amateur in any event, so only a few of the simpler types are shown in the photographs. Fig. 69 shows the two halves of a pattern similar to that

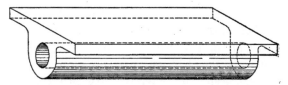

Fig. 86—Appearance of casting showing cored hole

of Fig. 66B, except that two sets of core prints, large and small, are used at P, P. No core boxes are shown, for standard cores were employed at the foundry. B and D, Fig. 69, are instances where both vertical and horizontal core prints occur in the same pattern; X denotes the vertical print. At C is a more complicated core box which goes with the pattern shown at Fig. 62B, and half of which may also be seen at H, Fig. 69. The method of moulding this pattern would be exactly the same as the operations shown in moulding the meter case, Figs. 79, 80, 81; the core box is more complicated because the interior of the casting must contain ribs and bosses. The disc-like core print may be seen on the face of the pattern at H. The outside of this core box was made round merely for convenience, and as far as the molding is concerned, could just as well have been square. The inside shape of the box is all that should be considered from the molding standpoint, for here the core is formed, and when it is placed upright in the mold (in this case) the hot metal surrounds it and leaves an opening in the casting the same shape as the core.

Another type of core can be considered; one which has to be placed below the parting line of the mold and which may be either horizontal or inclined. Fig. 68, J, and Figs. 83, 84, 85, 86 illustrate such a case. This is a one-piece pattern which will be molded with its flat side resting upon the board; therefore the parting line of the mold would come even with this side. The part of the casting that is to have the horizontal hole through it will be down in the drag, and it is obvious that if regular horizontal cylindrical core prints were used the pattern could not be drawn, for then the core print projections would tear up the sand. In a case of this kind, therefore, *the core prints must extend up to the top of the mold so that the pattern may be removed therefrom.* They may be given any convenient shape as long as they will draw; for instance, in this pattern they are an extension of the lower part and extend up even with the top and, therefore, even with the parting line of the mold. The body of the core, as previously stated, must be of the same diameter as the required hole through the casting. A core of this kind always requires the building of a special core box; even though the body of the core were cylindrical, a standard core could not be used, for the two end portions which fit into the core print depressions must be molded and baked with the body. In this case, a half core box may be used; such a one is shown at G, Fig. 69, and at Fig. 85, and the appearance of the finished core in Fig. 84. If the two ends were not alike, a half box could not be used, for the two half cores would then not fit together. Fig. 86 shows how the casting would look; *note that the core print projections never appear on the casting; their function is only to form a support for the core in the mold.*

Space will not permit a further discussion of cores and core boxes, for they occur in an endless variety, and every one would require a special description, but the foregoing should give the reader an idea of the principles involved. Therefore we will proceed to explain a few small but important details such as "fillets," "loose pieces," and allowance for shrinkage and machine work.

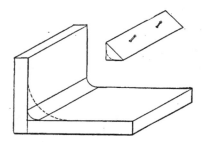

Fig. 87—The dotted lines indicate a fillet with too great a radius in proportion to the thickness of the pattern

A sharp corner should never be made on a pattern, whether it be curved or straight, but should be rounded off; and internal corners should be filled in by a curved portion called a "fillet." An exception to this rule would occur in the case of such corners as were formed upon that side of the pattern which would rest upon the molding board, as the top edge of the flange shown in Fig. 62A. Such a corner could not be rounded to any extent without interfering with the drawing of the pattern. A sharp corner always means a weak place in the casting, for strains are set up in the metal as it cools and are sometimes sufficient to cause a fracture, whereas a rounded corner, or one containing a fillet, causes a more uniform cooling stress in the casting. When cutting a pattern out of a single piece of wood, the fillets can generally be worked out from the solid, but if built up of separate pieces, sharp corners are usually formed and must be filled by a small strip of wood glued in and rounded with gouge and sandpaper after the glue is dry. This is illustrated in Fig. 87, which shows a triangular piece of wood with brads for temporarily holding it in place until the glue dries, after which the brads are drawn and the corner rounded. Strips of leather may be used for fillets instead of wood, especially in fitting rounded corners where wood fillets would be difficult to fit. They are fastened and shaped the same way as the wooden ones. Beeswax is sometimes used, the wax being softened and worked into the corner, but such work is not very permanent. A fillet should not be too thick in proportion to the thickness of the pattern, as in the dotted line of Fig. 87; this would be as bad as none at all, for the excess of metal in the corner would cause strains to be set up as the metal cooled.

"Loose pieces" are small parts of a pattern which project in such a way that if permanently fastened in place would not allow the pattern to be drawn from the mold. Instances of this are noted in Fig. 68, one being the lathe side M, which has a boss X on one side of the post, and the projection on the lower end of the base of the follow rest F. These parts are made separately and held in place by a long pin or brad. In molding, the sand is rammed down over the pattern until part of the loose piece is covered; then the pin is removed, leaving the piece in position while the remainder of the sand is filled in. When the pattern is removed, the loose piece stays in the mold and may be taken out without breaking up the sand. The piece must, of course, be small enough to be removed through the mold formed by the main part of the pattern. While the best practice calls for sliding dove-tailed joints and other more. or less complicated arrangements for loose pieces, the simple method of holding by a pin is perfectly satisfactory for small patterns.

Every metal shrinks to some extent upon cooling from a molten state. The amount of shrinkage depends on the size and shape of the piece as well as the co-efficient of expansion of the metal. For cast iron, the shrinkage allowance is generally one-eighth of an inch to the foot; this means that if a casting was required to be one foot long the pattern would have to be made twelve and one-eighth inches long, for the casting would shrink one-eighth of an inch in its length while cooling. But for most of the small patterns such as the amateur would build, unless very accurate castings were required, the shrinkage could be neglected.

Those parts of a pattern which correspond to the portions of the casting which are to be machine finished must be made thick enough to allow for the material removed from the surface of the casting while turning, boring or planing it. Usually about one-eighth of an inch has to be taken off in order to get down through the surface scale and give a smooth finish to the work. For instance, the top face, rim and lower face of hub of a casting of the face plate shown in Fig. 62A, would have to be turned off in the lathe and so the pattern would have to be made one-quarter of an inch greater in diameter than the finished size, the face one-eighth inch thicker and the hub one-eighth longer to allow for the machine work on the casting. If a hole was bored through the center, allowance would have to be made for removing at least one-eighth of an inch of metal from the sides of the hole while boring out, and more if there was a chance of the cored hole being crooked.

CHAPTER VIII

Explanation of the process—Description of a small plating outfit—Solutions used for the electro-deposition of copper, silver and nickel—Cleansing solutions for various metals—Polishing and finishing work.

The average amateur mechanic seems inclined to regard electro-plating as a very complicated and difficult process, involving the use of costly materials and apparatus. This is not the case, however, as the successful electro-deposition of copper, nickel and silver is a comparatively simple process, the practice of which easily comes within the resources and ability of the amateur. An electro-plating outfit should be included in the equipment of every workshop, and it is the purpose of the author to describe in the following lines the construction and manipulation of a small but practical outfit, which will enable the mechanic to properly plate and finish his machine or instrument parts.

While it is not the purpose of the author to give a lengthy treatise on the theory of electro-deposition, a brief outline of the fundamental principles involved will be given, as an elementary understanding of the theory of any process or operation about to be performed invariably proves helpful and advantageous in actual practice.

Pure water is a very poor conductor of the electric current, but if a small quantity of table salt (sodium chloride) is dissolved in it, it at once becomes a comparatively good conductor of electricity, and in this state is technically known as an electrolyte. If we immerse two electrodes in such a solution and pass a current be-

126

tween them, it will tend to decompose the table salt into its constituent elements, i.e., sodium and chlorine. The atoms of sodium will accumulate at the negative elec-

Fig. 88—A glass storage battery jar used as an electro-plating vat

trode or cathode and the chlorine will be attracted by the positive electrode or anode.

This is exactly what happens in the process of electro-plating. For an illustration, we will assume that we are plating with nickel. In place of the sodium chloride, nickel-ammonium sulphate would be dissolved in the water, and upon passing an electric current through such a solution we would find that the negative electrode would soon become covered with a thin deposit of metallic nickel owing to the decomposing action of the current. If we take the negative electrode out and substitute it with articles to be plated, we will find that the articles will undergo the same process and a deposit of nickel will form on them. If we wish to plate with copper, copper sulphate should be substituted for the nickel-ammonium sulphate, etc.

In many cases, where there are but a few small articles to be electro-plated, a small vat may be used with entire success and the results will be found to be just as good as those attainable by means of larger vats and more costly apparatus. For general workshop use, the outfit described in the following lines will be found to be both practical and serviceable.

The vat proper should be a square earthenware or glass jar with dimensions not smaller than 8 inches square and 10 inches deep. A glass storage battery jar of the proper size is very suitable. The top of the vat should be equipped with three $\frac{5}{32}$-inch brass rods, as shown in Fig. 88. One end of each rod should be threaded to receive $\frac{5}{32}$ machine nuts. Two of the rods are to be used to hold the electrodes, and the articles to be plated are suspended from the third one. On account of the necessity of varying the distance between the electrodes for different classes of work, it will not be found desirable to construct a permanent arrangement to hold the rods in place, as they are heavy enough, when equipped with the electrodes, to remain in any position they are placed in.

The electrodes of a plating vat should be of the same metal that is to be deposited. Thus, if we desire to copper plate, copper electrodes should be used; if we wish to nickel plate, nickel electrodes should be used, etc. For this reason, it will be found necessary to construct three sets of electrodes for use with the vat; one set of copper, one of nickel, and one of silver. Owing to the greater expense of silver, the electrodes of this metal are made much smaller than those of nickel and copper, and on this account it will be found necessary to plate one article at a time when depositing silver. As the silver plating solution is equally expensive, the amount prepared should only equal one-third that of the copper or nickel solutions, and the silver electrodes should be suspended into the solution by means of longer strips than those

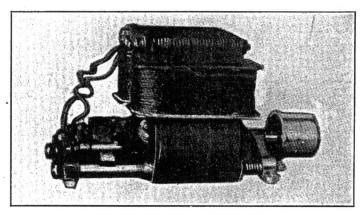

Fig. 89—An electro-plating dynamo

used on the other electrodes so they will be completely immersed. The dimensions of the various electrodes and the method of suspending them from the brass rods are plainly shown in the sketch.

While three Bunsen cells connected in series will be found to produce sufficient current to successfully operate the vat described above, the use of a small 8-ampere, 10-volt dynamo of the shunt-wound variety is to be recommended on account of the steady and unvarying current it is capable of generating. It is utterly impossible to use dry cells, as they polarize too rapidly for work of this nature. A small rheostat should be included in the outfit, as it is often necessary that the current be properly proportioned for the work required from it.

Successful electro-plating depends, to a great extent, upon the chemical purity of the solutions, and only the purest substances should be used in their preparation. If it is impossible to obtain pure distilled water, the next best substitute is rain water.

Solution for Copper Plating.—First, make a saturated solution of copper sulphate (blue vitriol) by dissolving the crystals in a gallon of pure water until it is found that the crystals will no longer dissolve. The solution is then said to be "saturated." To this preparation add about a half teacup of chemically pure sulphuric acid, care being taken that the acid is poured in a small, gentle stream. After filtering through blotting paper, the solution is ready to be used or stored away in flasks until it is desired to use it.

Solution for Nickel Plating.—Dissolve one pound of nickel-ammonium sulphate in one gallon of water. To this add about 2 tablespoonfuls of pure sulphuric acid. The preparation is then filtered, after which it is ready for use. After this solution is used for some time as a bath, it is advisable to test it with litmus paper to ascertain whether it is acid or alkaline, as there is a tendency for ammonium (NH4) to form, which renders the bath alkaline. In this case, sulphuric acid should again be added until the bath is just acid.

Solution for Silver Plating.—Obtain two ounces of silver nitrate from a chemical supply house and dissolve it in two quarts of pure warm water. To this add a solution of cyanide of potassium (a deadly poison), which will cause the silver to precipitate as crystals of silver cyanide. Immediately discontinue adding the potassium cyanide after it is found that all the silver has precipitated. The whole solution is then filtered through blotting paper to recover the crystals of silver cyanide formed. The filtrate may then be disposed of, as it is of no further use. Now place the silver cyanide crystals in a vessel containing about one and one-half quarts of pure water and to this add potassium cyanide, stirring the solution at the same time until all the silver cyanide crystals have dissolved. We then have a standard solution of the double cyanide of silver which is used in

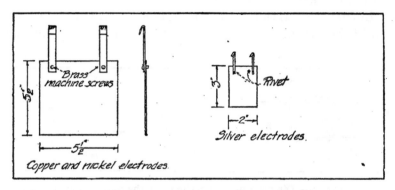

Fig. 90—Copper and silver electrodes for the electro-plating vat

silver-plating. The solution may be kept in a clean, stoppered bottle until it is used.

Cleansing .Solutions. — One of the most important operations in the process of electro-plating is that of properly preparing the surface of the articles to receive the deposit. The smallest speck of foreign matter upon the surface of the article to be plated is sufficient to cause the deposit to peel off. Many times the mere touching of the surface with the fingers so contaminates the object that it becomes impossible to electro-plate it successfully without again putting it through the cleansing process.

It is, of course, understood that the surfaces of the article to be plated should first be rendered sufficiently smooth by a mechanical process, if it is not already so.

In the case of the amateur, this can usually be accomplished by polishing the surface with fine emery cloth.

After the surface is mechanically prepared, it then becomes necessary to render it chemically clean before it is immersed in the plating bath. As it is impossible to prepare one cleansing or pickling bath that will be suited for all metals, it will be found necessary to mix several different pickles, one for each different metal.

Before the articles are immersed in the pickle, they should be dipped in clean water, and after they are

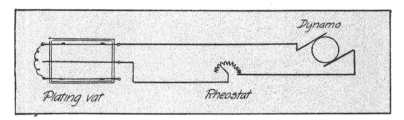

Fig. 91—Diagram of connections for the plating outfit ·

brought out of the pickle they should again be thoroughly rinsed in clean, running water before they are finally ·placed in the plating bath. The articles should be dipped in the pickle by means of a copper wire.

Pickle for Copper, Brass and German Silver.—One hundred parts of sulphuric acid, 50 parts of nitric acid and 1 part of table salt. Permit the preparation to stand one day before using. Use 100 parts of water and 10 parts of sulphuric acid for zinc.

Pickle for Iron and Steel.—One part sulphuric acid, 15 parts water, ½ part nitric acid. It is advisable to add a few pieces of zinc to such a solution.

When using the acid dips, especially in the case of copper and brass, care should be taken that the metal is not left too long in the pickle, as the acids act quickly and will pit the surface if permitted to act long enough.

The proper method is to alternately dip the articles in water and then in the pickle until the surface appears bright and clean.

If it is desired to plate a brass article that has already a fine polish upon its surface, the acid cleansing bath should not be used, owing to its tendency to destroy the polish. A dip composed of 1 part of potassium cyanide to 10 parts of water may be substituted for the acid dip in this case. It will be found necessary to leave the brass much longer in this than in the acid pickle.

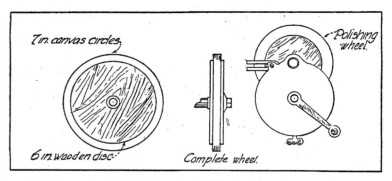

Fig. 92—A simple polishing wheel arranged on a small hand grinder

Hints.—When plating objects that have large projections, the electrodes of the vat should be placed as far apart as possible, otherwise an unequal deposit will result owing to the great current density at the projections where the resistance of the bath is least.

In electro-plating, care should be taken that the deposit does not form too rapidly, as plating of this nature invariably proves to have poor adhesive qualities and soon peels off. Equally wrong is the practice of permitting the deposit to form too slowly. The current should be regulated by means of the rheostat, until the deposit formed is flesh-pink in the case of copper, and milky white in the case of silver and nickel. If the current is not of the proper proportion, the deposit has a noticeable tendency to become dark in color.

If the plating solutions are not in use, they should be kept in stoppered vessels, otherwise they will become contaminated with foreign matter from the atmosphere.

The articles should be dipped in the pickle on a copper wire, bent in the form of a hook, and, immediately after cleaning, the articles should be placed in the plating vat by hanging the copper wire or hook on the central brass rod.

The fact that the amount of current passing through the bath is dependent upon the proximity and size of the electrodes should be kept in mind. If a single small article is being plated, it is necessary to move the electrodes closer together in order to reduce resistance and permit sufficient current to pass owing to the small surface of the negative electrode which is formed by the article to receive the plating.

Polishing and Finishing.—After the articles are taken from the plating vat, they should be washed in hot water and dried in a box of sawdust. They are then ready to receive the final polishing, which may be done on a small grinding head equipped with a buffing wheel. If the mechanic is not fortunate enough to have one of these in his workshop equipment, a good substitute will be found in a bench grinder, which may be fitted with a polishing wheel, as shown in Fig. 92. The polishing wheel is made by cutting out about twenty 7-inch circles from thin canvas and clamping them between two wooden discs as shown. The purpose of the wooden discs is to hold the canvas circles in place under the pressure of polishing. With a helper to turn the grinder and a little rouge on the buffing wheel, very good polishing can be done.

CHAPTER IX

Description of the engine—Procedure in machining and finishing the various parts.

This model represents what is probably the acme of simplicity in steam engine construction. It ranks favorably with the old type oscillating cylinder machines in this particular, although it is admittedly superior in point of mechanical efficiency and design. The entire absence of cross head and usual connecting rod, with their attendant difficulties from the builder's standpoint, leaves nothing to be desired.

The principle of operation may not be clear at first glance at the drawing. A rather comprehensive explanation will, therefore, be given. The motion of the piston rod is imparted to the slide crank or cross piece which travels up and down. Sliding within the hollowed out cross piece is a short cylinder which engages a pin on the crank disc attached to the shaft. The pin passes through the opening which runs the length of the cross piece.

As the piston rod moves up and down, it causes the crank pin to describe a circle, the cylinder engaging the pin sliding from one end of the cross piece to the other.

The steam ports are covered with a slide valve of the usual type and in all other respects the little engine represents standard model design.

The cylinder casting is best held in a three jaw universal chuck for boring and facing one end. It is then reversed in the chuck with the faced-off end backing up against the chuck face to insure that the end to be faced

135

will be truly parallel with the finished one. The valve seat is next to be considered, and the best way to machine this is to mount the bored and faced casting upon an angle plate, taking the necessary cut off the valve seat. If no angle plate is available, the facing may possibly be done with a file providing the worker is sufficiently expert in the use of this tool. A plane surface

Fig. 93—The slide-crank engine

is best produced by drawing the casting across the face of a large, fine file, rather than by attempting to run the file across the casting.

The cylinder heads may next be machined by holding the casting in the chuck by means of a chucking piece left for the purpose. Before removing the casting from the chuck, it is well to scribe the circle around which the holes for the cylinder head screws are to be drilled. By doing this in the chuck with a pointed tool, the layout will be perfectly coincident with the bore of the cylinder. It is afterward a simple matter to lay off and prick punch the locations for the six holes in each cylinder head. The hole for the piston rod and packing gland is, of course, to be drilled in the lower cylinder head casting before taking the latter from the chuck.

When laying out the holes for the supporting columns at either side of the engine, it is well to clamp the base plate and lower cylinder head together and to drill holes through both at one operation. This will insure perfect alignment. The columns are, as will be noted from the

Fig. 94—The complete set of castings for the slide-crank engine

drawing, lengths of cold rolled steel rod shouldered down and threaded at either end.

The hole in the center of the base plate may be spotted while the castings are still clamped together. It is, of course, quite essential that these holes line up perfectly to insure smooth running without undue friction.

The machine work on piston, slide valve, and eccentric is so obvious as to require no special explanation. The cross slide may require a word or two, however. The casting is made with the cross member solid, and it should be drilled ¼ inch for the sliding cylinder while held in the chuck. The casting may then be turned 90 degrees with the chuck jaws gripping the smaller projection while the tailstock center engages a center hole

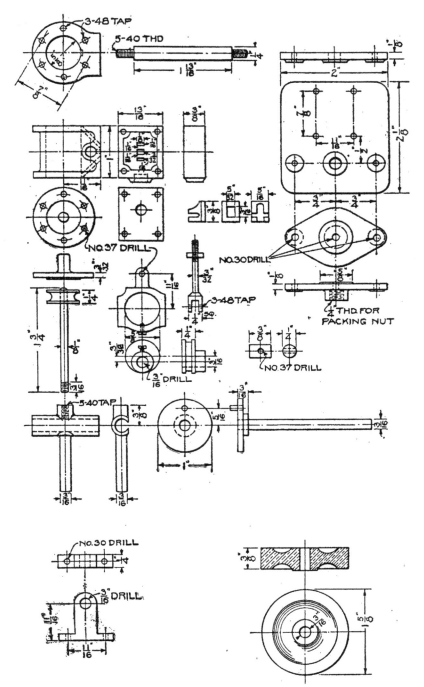

Fig. 95—Detail drawing of the engine

in the longer end. A steady, smooth cut is then taken along this longer projection, turning it down to a finished diameter of ³⁄₁₆ inch. Reversing the casting again, and using a sleeve to protect the now finished longer portion from injury by the chuck jaws, the smaller end may be drilled and tapped for the piston rod. Removing the casting, the builder may then proceed to file away the side of the cross member so that the slide crank pin may enter to link the sliding cylinder with the crank disc.

The bearing standards may next be drilled for the shaft with a drill a shade under ³⁄₁₆ inch and holes placed in the feet. After cleaning up the under side with a file to make it square with the perpendicular, the bearings may be mounted on the base plate, great care being taken to see that the holes for the shaft line up perfectly with a line scribed the length of the base plate and passing exactly through its center. This is easily done by passing a short piece of rod through the holes and in this manner sighting for alignment. When the holding down screws, have been inserted, a ³⁄₁₆ inch reamer may be passed through both standards, clearing out the bearings and making absolutely certain that they line up perfectly.

The turning of the flywheel finishes the machine work when the engine may be assembled.

CHAPTER X

A MODEL TWIN-CYLINDER ENGINE

Description of the type of the engine—Machining the cylinder castings, crankcase, valve chest, crankshaft and valve mechanism—Finishing the engine.

The model described below resembles the Westinghouse high speed stationary steam engines used in driving electric generators for lighting circuits. It is simple in design, serviceable in operation and presents no great difficulties in construction.

This engine, when constructed from a set of magnalite castings, makes a wonderful power plant for a model

Fig. 96—The twin-cylinder engine

boat of from three to five feet long, and it delivers considerable speed and power. It will also be found to be a reliable and consistent runner, which is a valuable advantage.

In building up a set of these castings, the first thing

140

to do is to study the blue print so that the general appearance of the engine will become familiar.

Cylinder Castings.—The crankcase end with the six lugs or projections is first filed flat and true. If a small

Fig. 97—Main castings for the twin-cylinder engine

lap grinder is available, this job can be done accurately and quickly on such a machine. Otherwise, the model maker will have to resort to filing. After this operation, the cylinder casting can be mounted on the face plate and the upward portion faced off. The casting will now

Fig. 98—The completed crankshaft for the engine, with flywheel, pistons and eccentric mounted in place

have to be taken off the face plate so that the center lines can be scribed off for the centers of the cylinder bore. This is best done by filling in the cored hole with cardboard and marking the centers with a lead pencil. The casting must now be mounted on the face plate again and the center marks on the cardboard lined up with the back center of the lathe. When this is done, the cardboard filling can be taken out and the boring of the cylinders started. After finishing this, proceed in the same way to bore the other cylinder, after which both should be lapped out perfectly smooth. This is very essential

Fig. 99—Top view of the engine showing the steam chest uncovered

for a perfect running engine and for the development of maximum power.

Crankcase.—File the crankcase flat so that it makes a perfect joint with the cylinder casting. This job can also be performed on a lap grinder with success. Holes

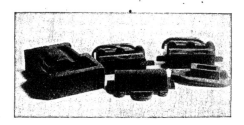

Fig. 100—Complete set of patterns for the twin-cylinder engine

are drilled and tapped and then both members are screwed together. The crankshaft bearings are best drilled by using the back center punch marks at each end, drilling first one side and then the other. A No. 1 drill should be used and then the hole carefully reamed out with a standard ¼-inch reamer.

Valve Chest.—This casting is filed or ground flat to make a perfect joint with the top of the cylinder casting. Necessary holes are drilled and tapped and the end slotted for the bell crank. The bell crank is the medium between the eccentric and the valve stem. The inside of the steam chest can be milled out if a milling machine is available. Otherwise, one must resort to the old reliable file and scraper. The valve stem passes through one end of the chest which has been drilled and fitted

Fig. 101—The crankcase of the engine after being machined

with the usual packing nut. The valve chest cover is merely a piece of ⅛-inch plate and screwed on the top of the valve chest. This plate has a hole drilled and tapped to take the steam pipe from the boiler.

Pistons.—These are put in the lathe and turned to a snug fit in the cylinders. It is understood that the pistons must work freely. The pistons have "V" grooves turned in them which act as water rings. For extreme high pressure, piston rings of steel must be fitted. Holes are drilled in the pistons to take the ⅛-inch wrist pin for the connecting rods.

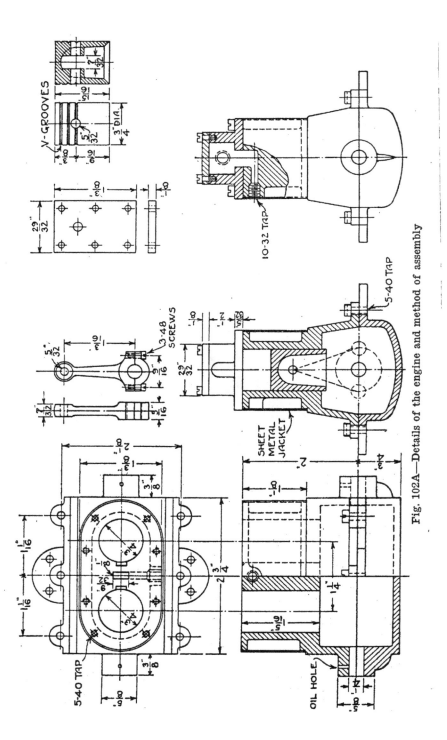

Fig. 102A—Details of the engine and method of assembly

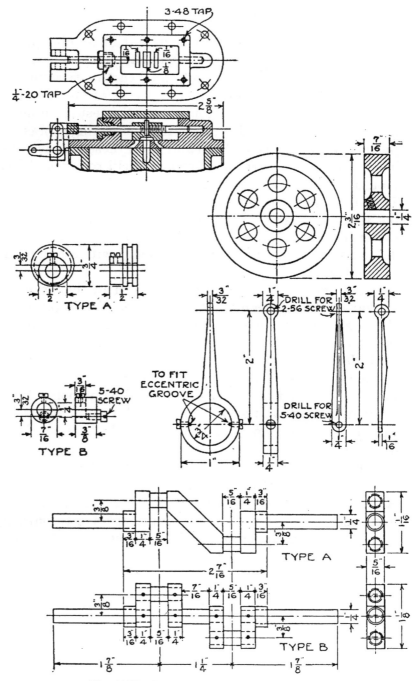

Fig. 102B—Details of the twin-cylinder engine

Crankshaft.—This can be made out of a solid piece, but the designer of this engine has had some experience with built-up crankshafts and can heartily recommend them. In fact, he prefers them, knowing of one instance where a built-up crankshaft has stood the test of running from 10,000 to 12,000 R.P.M. without breaking. Make a jig for the crank webs. Drill and ream these standard ¼ inch, and then procure some oversize drill rod (about 253 thousandths) and drive this through the crank webs. Then pin and braze the joints, after which the portion of the shaft that is between the web can be sawed out and the crankshaft cleaned and filed up smooth. This produces a very strong and rigid crankshaft with a minimum of trouble.

Both types of crankshaft are shown in the drawing so that the builder can make either one. While the solid shaft is more durable, it is much more difficult to make than the built-up type.

Valve Mechanism.—The drawing shows two different methods of making the valve mechanism. One is a lathe job, and the other can be made with ordinary tools. Both methods are good; the only thing in favor of one is its ease of construction, and this can easily be seen from drawing.

The only thing left to be done now is the putting on of the lagging, setting the valve and testing the engine. As for finish, a nice maroon enamel well baked on in an oven and then left to stand a couple of days to harden, gives à splendid appearance. The flywheel of the engine is merely a simple lathe job.

CHAPTER XI

A SINGLE-CYLINDER ENGINE

General procedure in machining the .engine parts, employing the most practical methods—Finishing the engine.

This particular design of engine is really the forerunner of the "Speedy" class of engines, having been built before the slide crank engine was designed. The engine has proven its worth on several occasions.

In the finishing of a set of these engine castings, the first thing to take in hand would be the 'cylinder casting. This can very easily be held in a 3-jawed chuck and bored out with a boring tool ⅝ inch diameter. After the boring is completed, the cylinder should be faced off before the casting is taken out of the chuck. This insures the bore and the bottom of the cylinder being absolutely square with one another. Now the casting can be taken out of the chuck and turned around, using a packing piece behind the cylinder and against the chuck face to get both ends parallel. The next operation is to drill and tap the six holes for the cylinder head and base. If the builder is in possession of an angle plate that can be bolted to the face plate of his lathe, the cylinder can be mounted so that he can face off the valve seat of the cylinder. Otherwise he will resort to the more tedious process of file and straight edge. In drilling the steam-chest parts; they must first be laid out (with a scribe) very accurately and then drilled with a hand drill to the depth shown on the drawing. The two steam ports are then drilled from their respective ends to meet the end ports drilled in valve face and the exhaust port is drilled to meet the center port. Care must be taken when drill-

147

ing the ports that the drill does not break through from one port to another.

The steam chest can now be fitted to the valve face after being filed or ground flat and true on both sides. A good method is to lay out the holes on the steam-chest casting and drill same and then use it as a jig for the holes to be drilled in the cylinder and also for the holes

Fig. 103—The upright engine assembled, ready to run

in the steam-chest corner. The various parts can now be screwed together, but not permanently.

The cylinder base casting can now be taken in hand. This has chucking pieces cast on so that the necessary turning can easily be done. When facing off the standard lugs, do not forget to face off the seat for the cross-head guide. This will save a lot of work afterward when the cross head guide is fitted. All holes should now be laid out and drilled as per drawing and then fitted to the cylinder. Now turn up the piston and make the piston rod with the cross head, fit the cross-head guide to cylinder base. Assemble all together and see that they work free. Do not forget to put the packing nut on the piston rod before screwing on the cross head. The slide-valve drawing is in detail and really needs no explanation. The

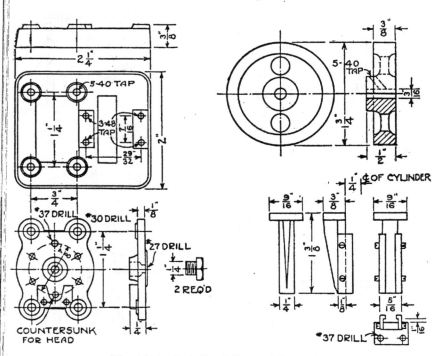

Fig. 104A—Details of the upright engine

eccentric casting will also be found to have chucking pieces cast on. First, put it in the chucks so that the 5/8-inch diameter can be turned, then take the offset and put in the chuck, center same and drill it with a 3/16-inch drill. You will find that the throw of the eccentric will be just right. The eccentric strap is then very easily made.

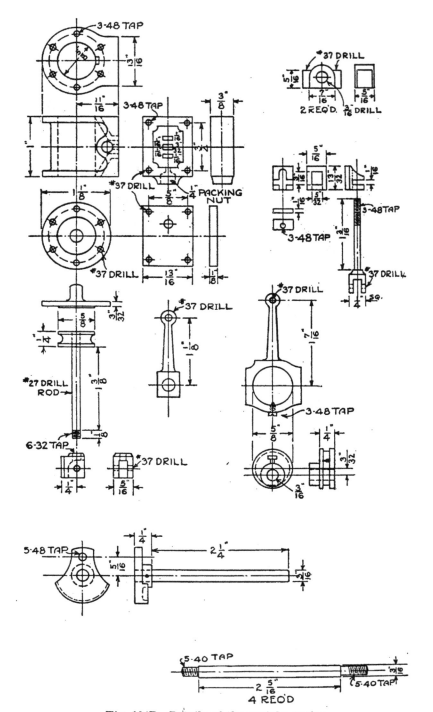

Fig. 104B—Details of the upright engine

The four $\frac{3}{16}$-inch steel standards are turned up at each end and threaded $\frac{5}{40}$, as per drawing.

In building the crankshaft, the crank disc is first ground flat, then drilled in the center for a driving fit for the $\frac{3}{16}$-inch shaft. After shaft has been driven in, it must be pinned and soldered to prevent it coming loose when running at high speed. Now the hole can be drilled and tapped for the shoulder screw which acts as the crankpin.

The engine base must first be ground or filed flat on the top and bottom. In laying out the holes, care must be taken that the upright standard holes match those on the cylinder base. If the builder would make himself a jig, he could use same for both the base and cylinder bottom, thus insuring absolute alignment of the four standards.

The holes for the crankshaft bearings are then drilled and tapped and the bearings fitted and drilled for the crankshaft. The flywheel needs no explanation, being a very simple lathe job.

If, in assembling the engine, the cylinder and its various parts, the engine base with the standards, crankshaft and eccentric are considered separate units, the builder will find that if he has these two units working properly it will be a very simple matter to put the cylinder unit on the four standards and connect up his valve motion and connecting rod. Now lag the cylinder with some Russian iron and enamel the cast parts red or green. The contrast between the enamel and the polished steel parts will make a very good looking model marine engine.

CHAPTER XII

Description of the engine and various parts—Lathe and machine work necessary in finishing the engine—Making a built-up crankshaft for the engine.

The beautifully executed model pictured and described in this chapter was built by the famous London model engineering firm of Whitney in City Road. The model is of brass throughout. Possibly for flash boiler steam, and that is admittedly the best for the purpose, it might be well to specify cast iron for the cylinders and pistons. It is said that brass or gunmetal will pit and otherwise cause trouble with very highly superheated steam. Accordingly, it is the prerogative of the builder to decide which material to use. He has but to weigh the ease of construction of brass with the superior operating qualities of cast iron.

While this model is not recommended as a project for the dabbler or the rank amateur in mechanics, its construction is delightfully simple for the worker who is possessed of a lathe and who knows how to use it. A screw cutting lathe is not essential as all of the work can be done with a small speed lathe fitted with a slide rest. Of course, the heavier the lathe, within limits, the easier the job and the better the workmanship. However, the advanced model maker who is capable of using his head to overcome difficulties need not hesitate to undertake the construction on even a small bench lathe.

The illustrations cover every detail of the engine, the working drawing being supplemented with photographs of the part of the machine ready for assembly.

It is assumed that the worker who attempts the construction will be sufficiently familiar with lathe practice to be able to work directly from the drawings without detailed instruction.

The one really difficult pattern is that for the standards supporting the cylinders which are cast *en bloc*.

Fig. 105—The twin-cylinder marine engine as it looks when finished

The dividing line is, of course, that through the stiffening web. A core is scarcely justified in the tunnel for the cross-head. The extra weight of metal, particularly if cast iron is used throughout, is not sufficient to warrant the extra trouble. Besides, it is practically as easy to drill through the cylindrical portion, following with the boring tool, as it is to get beneath the glass-hard scale left on the casting where the core has been.

The bed-plate, cylinders, flywheel, eccentrics and straps, bearing caps, steam chests, and the few pipe fittings which are not standard, constitute the bulk of the

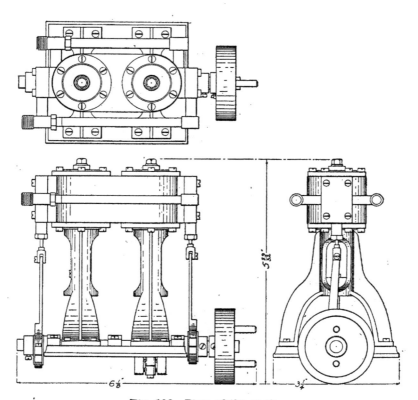

Fig. 106—Plan of the engine

other pattern work. The pistons, connecting rods, crankshaft and packing glands are about as easily worked out of rolled stock as from castings. The one exception is the cross-head which is something of a nuisance to machine. The blind slot in the cross-head is a difficult piece of work unless it is cored out in a casting and even then it is still a nuisance. Any worker who has tried to worry out a slot of this kind will appreciate the truth of the

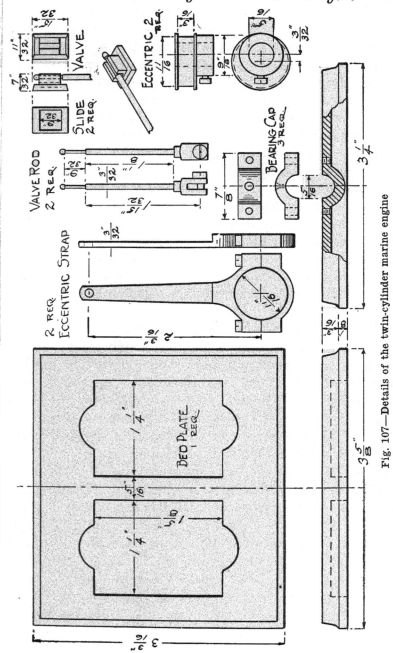

Fig. 107—Details of the twin-cylinder marine engine

statement. With good foundry work, the hole may be cored with more than fair success. Under such circumstances, the little pattern is certainly worth making.

The bed plate is a light casting that may be machined in one of several ways. Obviously, the professional method would be in a shaper, planer, or milling machine; but how many model makers can boast of one of the trio? Next in order comes a straight facing job on the face plate of the lathe. The bed plate casting can be secured readily to a wooden face plate screwed to the iron one. Centering the job by means of scribed diagonals enables the worker to center, and consequently balance, the casting on the face plate.

It is best to face off the under side first, then reversing the casting and facing off the upper side to which the engine standards are attached. The edges may well be merely cleaned up with grinder and file and afterward painted unless the worker prefers a job finished in bright metal only.

The crankshaft bearings are formed by drilling half into the bed plate and half into the bearing caps, one of which is shown above the part sectional view of the bed plate in end elevation. After cleaning up the bearing cap castings with file and grinder and finishing the under side by *drawing the casting over the surface* of a clean, fine, flat file, the caps may be located on the bed plate, the holes spotted, drilled and tapped for 6/32 fillister head screws, and the caps secured in place before the drilling for the crankshaft is attempted. Prick punch marks should be placed on each casting and just below it in the bed plate so that the bearing caps may be replaced in the same order after they have been removed.

The drilling should be done in the lathe. Build up a structure on the lathe bed to a height that will bring the dividing line between bearing caps and bed plate exactly in line with the lathe centers. Spot the hole to

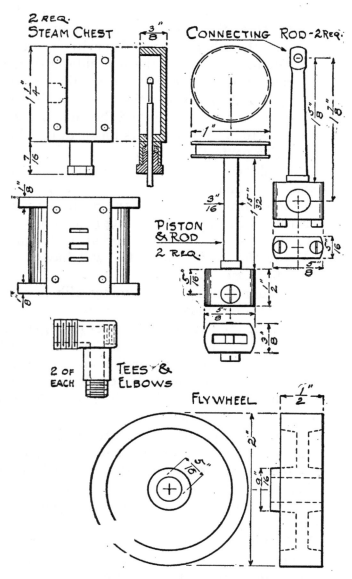

Fig. 108—Details of the twin-cylinder marine engine

be drilled in the casting on *each side* and start the drill on the one side while the casting is backed up by the tailstock center on the other. This will positively insure accuracy and alignment. When the one side is drilled, run the drill right through until it passes through the middle bearing, and then reverse the work so that the drilled side rests against the center and the undrilled side is presented to the tool. When this is finished, the three holes are sure to be perfectly in line.

The drill should be $\frac{1}{64}$th of an inch under $\frac{5}{16}$ if a reamer is available, and one is really necessary in this case. When the drilling is finished, the $\frac{5}{16}$-inch reamer is passed right through the three bearing holes which finishes them to size and, of course, to line also. To make a running fit for the shaft, the latter will have to be ground in with fine emery and oil.

The two cylinder castings are exactly alike, so the operations will be described for one only. A simple method of facing off the bottom of the feet is to grip the top, or lower cylinder head portion, in a three-jaw chuck, centering the central cylindrical portion which will ultimately be bored out. With the tailstock center brought up, the cut may be taken readily from the feet. The casting may then be taken from the chuck and mounted upon a face plate with the faced-off feet secured to it through the holes which will later take the final holding down screws when the engine is assembled. The end of the casting is then center-drilled, and the center brought up to permit the facing off of the cylinder head portion to dimensions given in the drawing.

When this is finished and a scribe circle struck with the sharp tool on the face of the cylinder head portion to serve as a guide in locating the holes for the screws, the tailstock center may be removed and a drill chuck substituted. A large drill may then be run through the casting to start the cross head tunnel. The greatest care must

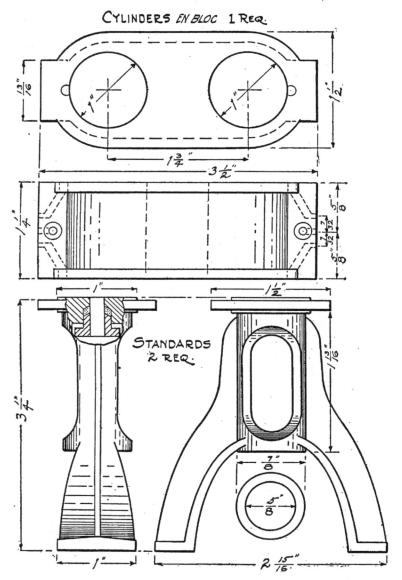

Fig. 109—The cylinder block and standards of the twin-cylinder
marine engine

be taken here to see that the drill does not bite and to this end it should be properly ground. If brass is used, the lip of the drill should be taken off by all means; with cast iron it is not so important but, even so, there should be no unnecessary risks taken to save time.

The boring may then be started through the drilled hole. If the depression on each side of the cylindrical

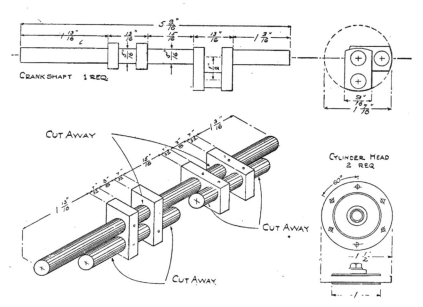

Fig. 110—The crankshaft of the twin-cylinder marine engine

portion catches during the operation of boring, it is very much better to use a steady rest on the turned cylinder head portion. This may be done without defacing the work seriously, and if it is marred, a light cut at the end will remove the marks. The steady rest will prevent danger from bites due to the lack of stiffness in the work.

To get a dead smooth cut at the final one, the boring tool should be sent through in the usual way with the cutting edge doing its chipping after the manner of a diamond point on external work; then, *and this is the trick,* without turning the cross feed handle, bring the tool back out again with the natural spring of the tool, taking a dragging cut after the fashion of a side facing tool. This final cut will leave the inside of the tunnel with a burnished surface if the feed has been very slow and the cut exceedingly light.

From the description just given, the worker will note that the end of the casting that would naturally form the cylinder head has been bored out. There is a double reason for this: First, there is no other easy way of boring the cross head tunnel and, secondly, it is much easier to fit a plug containing the packing gland, the construction of which is well shown in the drawing, to the bored out casting than it would be to attempt such a fitting up inside the tunnel. The actual packing is a piece of lamp-wick well greased and wound around the piston rod.

The two cylinders are cast *en bloc.* The casting is considerably lighter than the pictures of the finished engine would seem to indicate as the smooth, external surface is of sheet metal which forms a "lagging" or covering which encloses an air space to keep the cylinders hot and in that manner prevent, as far as possible, the condensation of the steam within the cylinders. The drawings show how this space is formed, the dotted lines indicating the lines of the actual cylinder wall in the plan view.

For facing off, the casting may be mounted upon the wooden face plate with the casting centered. After the cut is taken, the work may be reversed to face the other side. The casting is then scribed accurately with center line and the two holes spotted to indicate cylinder bores. The casting is again mounted, but this time the center is brought up to one of the spot marks indicating one cylinder bore. After clamping securely, the casting may be

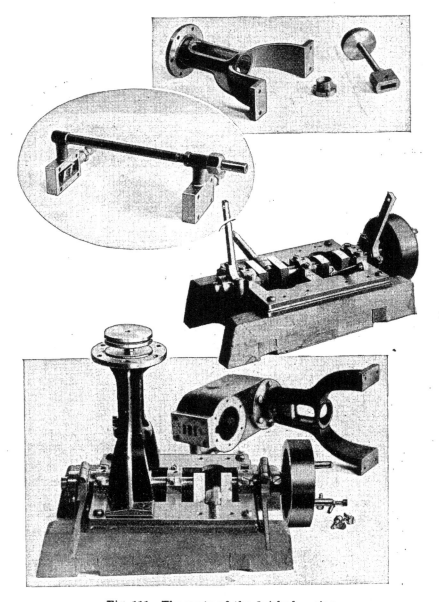

Fig. 111—The parts of the finished engine

drilled, but first of all the whole job should be balanced by bolting a small mass of metal to the side of the face plate to compensate for the lack of balance and to prevent vibration which would destroy the accuracy of the work. The boring of one cylinder finished, the casting may be changed over so that the other may be bored.

The casting may then be removed from the face plate and the ports drilled and chipped out. Before doing this operation, however, the slide valve faces may be machined. The right way to do this is on an angle plate in the lathe after the cylinders have been bored and faced off. An alternative method is to file and grind the face true, or, as nearly so as the skill of the worker will permit.

The little kink used for the slide valve rod is a good one, and the isometric sketch will aid in showing its principle. The reader will note that this stunt affords a universal coupling within the steam chest, making it possible for the pressure of the steam to keep the slide valve tight against the face.

The piston is turned right on the piston rod which serves as an arbor. This insures accuracy in the finished job. The piston is packed with wicking in the groove turned for it.

The crankshaft requires special attention as it presents some difficulties to the uninitiated. Two methods of construction are permissible. One is the solid forging construction and the other the built-up shaft method. The latter only will be considered here as it is believed the former is well understood by those who are capable of handling it.

The crankshaft may be made with crank webs at 90 degrees or 180 degrees. The model pictured has the 90-degree crankshaft which presents the advantage of no dead center. However, such a shaft is unbalanced and when the engine runs at high speed the vibration is rather excessive. The 180-degree shaft obviates the latter difficulty but introduces the unfortunate dead center when the two pistons are at the extremes of their travel. However, it is for the builder to decide which of the two evils is the lesser. To beat them both, he has only to add a balancing weight to the crank web opposite the pin.

The built-up crankshaft is constructed of cold rolled steel for the webs and steel drill rod for the shaft and pins. The webs are laid out, centers punched, and holes drilled $\frac{1}{64}$ inch under $\frac{5}{16}$. The drilling may be done in the lathe, the work being held against the tailstock drill pad and the drill in the lathe chuck. For convenience in handling, it is well to drill all holes in the bar of steel before it is cut to form the webs. The holes must be very carefully laid out and center punched.

When the drilling has been finished, the bar may be cut up into the four pieces forming the webs. These pieces may be stacked and a piece of cold rolled steel passed through one set of holes to line up the pieces. A clamp may then be put on and a $\frac{5}{16}$-inch reamer run through the uncovered set of holes. The latter may then be filled with a second piece of rod made for an easy push fit. The first rod is then removed and the reamer run through that set of holes. The webs are then to be assembled on an arbor, placed first in one set of holes and then the other, so that a lathe cut may be taken off the ends. This will finish the webs nicely.

The final assembly should be upon the $\frac{5}{16}$-inch drill rod forming the shaft and wrist pins. This rod will be a very snug fit in the holes, and before the shaft is assembled the web pieces should be removed from the arbors and heated, on a piece of wire, in a Bunsen flame until they just begin to turn blue. *While hot,* they are placed, by means of tongs, on the cold shafts in exactly the positions they are to occupy. The shafts should previously have been marked plainly for each web. While the webs are hot, they will be an easy twist fit on the shafting; when they cool, however, they will contract upon the shafting so that it will be all but impossible to turn the shaft rod in the hole.

When the job has cooled, holes may be drilled through each web and through the shaft it encloses for small steel pins which will finish the job of locking the parts together. After these pins have been driven, and not before, the superfluous shafting may be cut away, leaving the complete crankshaft ready to be finished up either between lathe centers or with a fine file. The cleaning up operation is merely one of removing the burrs of the hacksaw used to cut away the extra shafting from between webs.

This completes the machine work of importance. The cylinder heads, fly-wheel, exhaust and inlet pipes, elbows, and tees are all simple pieces of work that need not be described.

CHAPTER XIII

How flash steam plants operate—Description of the various parts and
fitting employed in a flash plant—Regulation of the water supply—
Lubrication—Difficulties in adjustment—Gasolene burners for flash
steam plants.

Flash steam plants are more adaptable to the propulsion of model steam boats than ordinary "pot" boilers, as they generate steam more rapidly and are therefore able to furnish more power to the engine. Although somewhat more unwieldy to handle than ordinary boilers and somewhat difficult to adjust, they are, nevertheless, preferred for model speed-boat work. All the present records are held by boats propelled with flash steam plants, which alone is enough to indorse their use.

The average American model maker is not very familiar with the operation and construction of flash steam plants and the following paragraphs are devoted to the method of operation and the general features of construction.

The illustration, Fig. 112, shows a complete flash steam plant and the method of connecting the various parts. The small tank shown at A is used to hold the gasolene which is fed to the burner C. The gasolene passes through the vaporizing coils, which are wound around the outer surface of the burner. The nipple at the end of the vaporizing coils has a very small aperture through which a fine spray of gasolene passes into the cylinder of the burner, where violent combustion takes place, and the flame produced shoots forward into the

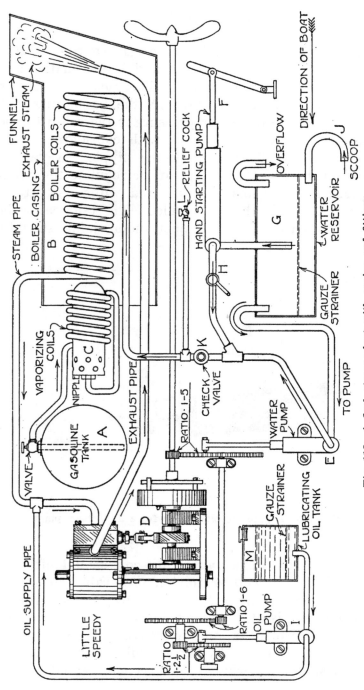

Fig. 112—A flash steam plant with engine and fittings

boiler coils, which contain the water. The gasolene burner is very simple in construction and is automatic in its operation after being started. The gasolene from the fuel tank, in circulating through the vaporizing coils, becomes hot and its vapor pressure is increased to such an extent that it comes forth with considerable force through the small opening in the nipple. A valve is placed in the fuel tank to cut off or regulate the supply of gasolene. The small opening and cap is also fitted to the tank through which it is filled. The tank can either be made of sheet brass or copper and all joints should be

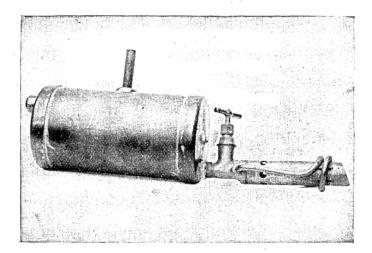

Fig. 113—A gasolene blow torch for a flash-steam plant

silver soldered. The vaporizing coil should either be made of steel or copper tubing, preferably steel. The burner is started in much the same way as the ordinary gasolene torch is lit. The valve in the fuel tank is opened and a little gasolene is run into the burner. This is ignited and at first a very smoky flame results, but as this flame continues, the vaporizing coils become heated and the flame gradually changes to an intense blue and produces a great amount of heat. After the burner is operating satisfactorily, the feed valve on the fuel tank is opened full. After a few minutes of operation, the engine will be ready to start and it is given a few sharp turns to start the main water pump, shown at E in the drawing. The gasolene burner can also be heated by means of another burner to start it.

The water pump E is generally geared to the propeller shaft with a ratio of 5 : 1, but this may vary with the stroke and bore of the pump. It is the purpose of this pump to start and maintain the circulation of water through the boiler coils. A hand starting pump is also included in the equipment, and this is shown in the drawing at F. If the equipment is provided with a hand starting pump, it will not be necessary to twist the engine shaft to start the main water pump E, as the initial pressure of water can be raised by the hand pump, thereby obviating the necessity of turning the engine over, which, in some cases, is rather a dangerous thing to do, as the engine is apt to start off and catch the operator's fingers in the mechanism. It must be remembered that the engine starts with a rush and has considerable force behind it. When the hand pump is used in starting, the engine is placed at its dead center. It will be seen from the drawing that both the hand pump F and the main water pump E obtain their supply of water from the reservoir G. The water is taken from this source and forced through the boiler coils. The water is only able to pass through the main pump E in one direction, as the valves prevent it from flowing backward.

When the water from the pumps reaches the boiler coils, through the center of which the gasolene flame is burning, it instantly "flashes" into steam and by the time this steam has reached the end of the coils toward the engine, it has become highly superheated and imparts

maximum power to the piston of the engine. When the
engine is starting, it is sometimes advisable to manipu-
late the hand pump, as this assists the main water pump
to pick up and produce maximum pressure in the water
flow.   After the engine is working satisfactorily, the
small hand pump F is cut off from the rest of the system
by means of à valve shown at H.   The engine is now
able to drive the water pump E, which is powerful enough
to produce the necessary flow of water in the boiler coils.
The engine is also required to drive the small oil pump
which is shown at I.   This pump is supplied with lubri-

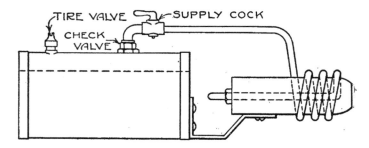

Fig. 114—Drawing of a gasolene torch for a flash-steam boiler

cating oil through the reservoir M.   The gear ratios of
the pump gears are shown in figures on the drawing,
and it will be noticed that the oscillating motion of the
pump piston will be very slow.   Owing to the fact that a
small amount of solid matter contained in the oil would
prove fatal to the successful operation of the engine, a
copper gauze strainer is used in the tank and the oil must
pass through this before it reaches the feed pipe at the
bottom of the tank.   Solid matter is thus prevented from
fouling the engine.   The oil pump produces a discharge
of oil in the steam supply pipe from the boiler coils.   The
lubricating oil is thus mixed with superheated steam just
before it enters the cylinder of the engine.   The lubri-

cating oil used should be rather viscid so that the high temperature of the steam will not cause it to become too thin and burn.

Attention is again drawn to the water reservoir. The capacity of this is not necessarily large; on the contrary, it is quite small. It will be noticed also that the water is first filtered through a copper gauze strainer before it reaches the feed pipe of the pumps. This prevents trouble that would be caused by solid matter in the water reaching the pumps. Due to the fact that the water reservoir

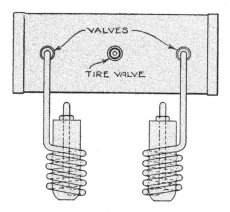

Fig. 115—A double gasolene burner

is placed below the water line of the boat, it will remain full when the model is at rest. When the boat is in motion the water reservoir is kept full by the scoop shown at J. The rapid forward motion of the boat tends to force water upward through the scoop into the tank, thereby keeping it full. A small overflow pipe is arranged at the top of the tank and this discharges over the side of the boat. The overflow pipe also provides an escape for air when the boat is first placed in the water.

A covering is made for the boiler coils and burner and this is usually shaped out of steel and prevents draft from deflecting the flame from the center of the boiler coils. The steam exhaust from the engine cylinder is led to the funnel on the covering of the boiler coils, where it discharges into the outer atmosphere. This tends to produce the proper circulation of air in the covering, so that the burner obtains a sufficient supply of oxygen for its most efficient operation. The check valve in the water supply pipe shown at K prevents water from returning if the delivery valve in the main pump does not function properly. The relief cock shown at R is opened when it is desired to stop the plant. When the relief cock is open, the water and steam pressure of the system is relieved. All the joints, cocks and valves of the water system must be absolutely air and water tight to produce an efficient plant, as the least leakage will seriously impair the successful operation of the device, as the pressure is very high throughout the entire system. This precaution will not be necessary with the oil-feed system, as the oil used is very heavy and no trouble will be experienced by leaky joints if ordinary care is exercised in making them. The water pumps for use with flash steam systems are always provided with mushroom valves. Such valves are more reliable than the other type, and their use is strongly recommended.

After a flash steam plant is started, it will work automatically, providing all the parts are in good running order. The blow-lamp heats the boiler coils and these in turn generate superheated steam which is fed directly into the engine cylinders.

Flash steam plants, however, are very difficult to get to the proper adjustment, and when once adjusted are very easily put out of adjustment by minor causes. Being that every square inch of surface in the flash coils is heating surface, the amount of water supplied to the boiler must be exactly what it requires. The heat must also be regulated so that the temperature of the steam will be of just the correct value for the engine's needs. Being that the steam is highly superheated before it enters the engine, the valve and engine parts must all be made of steel to withstand the severe attack of the heat. An increase of heat causes the temperature of the steam to rise and also increases its expansion. Many times the temperature of the superheated steam is so high that it will burn up the lubricating oil before it reaches the cylinder of the engine. When the lubricant becomes ineffective the results are apt to be very disastrous to the engine, as great friction is caused and seizing is apt to occur.

If the heat of the burner falls off for any reason, the steam is not raised to a sufficiently high temperature and will not be thoroughly dry at the time it enters the cylinder of the engine. When this condition occurs, the boiler coils are unable to vaporize the water that is circulating through them from the pump, and in a short time the boiler floods and the engine is fed with a large percentage of water in place of steam. It is necessary that a flash steam plant, to operate successfully, must be supplied with exactly the amount of water, heat and steam it requires in operation. Thus, it will be seen that to obtain the maximum power from such a plant adjustments must be made very accurately and with great care.

For model power boats, flash steam plants are much more successful and efficient than ordinary pot boilers, as they generate greater power for given weight and fuel consumption.

It will be understood that it is almost impossible to heat a flash steam boiler by any method other than the gasolene blow torch, as the flame from this completely covers the surface of the entire boiler coils.

Flash boilers can also be made with double coils and in this case it is, of course, necessary to employ twin burners to fire them with. Double coil flash boilers are capable of developing considerable power and are very suitable for racing craft of the larger type.

It is necessary to encase all flash boilers with Russian iron which prevents heat radiation and also protects the flame from drafts. The inside of the Russian iron casing should be lined with asbestos.

CHAPTER XIV

A description of the engine and what it is capable of doing—Machine work necessary to finish the engine—A flash steam plant for the engine and how to make it.

In view of the great activity recently evinced in the designing of model power plants for aviation and other purposes, this compact and efficient power plant should prove of considerable interest. Providing as it does a self-contained plant of sufficient power to make it emi-

Fig. 116—The four-cylinder steam engine assembled

nently practical and of great utility, the construction of this engine at once supplies the model maker with a most interesting project, and furnishes him with a reliable source of power. While no special effort was made to secure lightness, as the engine was originally designed to propel a canoe, for which purpose it has proven perfectly successful, the power plant is not too

175

heavy for the large airplane' model which it would be capable of driving, and would undoubtedly prove an ideal solution to the propulsion problem in connection with such a model. In fact, the author considers that it would vindicate the employment of steam vs. other prime movers in whatever field in which it might be put, and in addition, the construction alone would be of sufficient

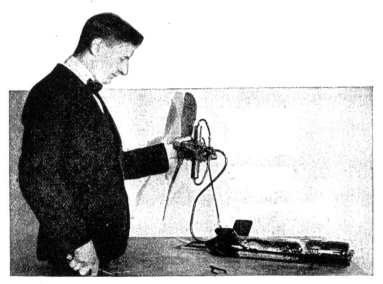

Fig. 117—The four-cylinder engine and its flash boiler

interest to repay the model maker for his expense and labor.

The engine, which will deliver a maximum of about 2½ H.P., utilizes steam from a flash boiler in four single-acting, radially disposed cylinders. From the standpoint of compactness, there is little doubt that this arrangement is superior to any other, and a great simplification of valve mechanism is also secured. In this case, the distribution of steam to the four stationary cylinders which are, of course, at an angle of 90 degrees to each other, is

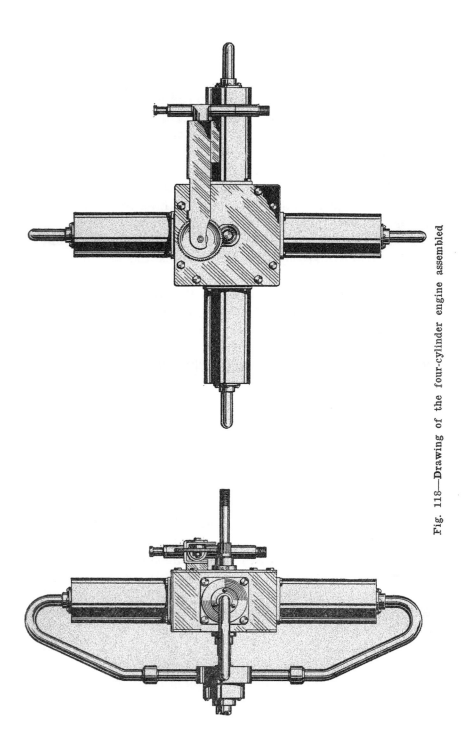

Fig. 118—Drawing of the four-cylinder engine assembled

effected by means of a rotating member flexibly coupled to the crankshaft and revolving in a casing connected with the steam supply and with the cylinder heads. This member carries on its periphery two ports, one, communicating continually with a small chamber under pressure from the boiler, serves to admit steam to the cylinders in succession as it passes the ends of the pipes connected with the cylinders, and is of such length as to cut off the admission of steam to the cylinders after about 20 degrees of the inlet stroke; the other port is in communication with the atmosphere by a series of holes and permits the cylinders to exhaust during about 80 degrees of the stroke. The general appearance of the engine is conveyed by Fig. 118. The four pistons are all connected to a single crankpin by means of a disc rigidly affixed to one connecting rod, and carrying supports for the other three rods on which they may pivot slightly as the crankshaft revolves.

One of the most unique features of the engine described below is that no castings are required; all parts being made either of cold rolled steel or brass stock.

The crankcase of the engine is constructed first. This particular part is shown clearly in the detail sketch, Fig. 122. In bending the stock, care should be exercised as this forms the foundation of the whole machine and any inaccuracy here would be fatal to the successful operation of the engine. The joint in the case is dovetailed and while this may appear difficult to accomplish, it is the only practical procedure. After the joint is accurately cut, the case is scalded in boiling water to remove all greasy substances from its surface. After this, the joint is silver soldered in a smokeless forge fire.

The bosses, which hold the cover plates of the crankcase to the case proper, are riveted to the sides as shown in Fig. 122. After the cover plate is drilled, it can be held in place on the case and used as a jig to drill the holes in the bosses. A ³⁄₃₂ drill is used and the holes tapped out with a ⅛-40 tap.

The holes in the crankcase, over which the cylinders set, are bored out on a lathe as it is impossible to do this on a drill press owing to the light weight of the stock. The main bearings are bronze, split and held in place with a flange and nut. The bearings should be carefully reamed out with a hand reamer.

The crankshaft is worthy of mention on account of the difficulties involved in making it from a single piece of stock. A piece of stock measuring ½ x 1¾ x 6 inches

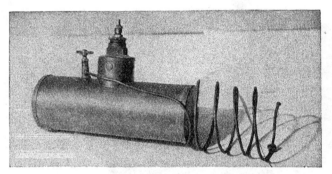

Fig. 119—The fuel tank for the gasolene burner

will be needed. After cutting out the rough form with a hack saw, the main shaft is turned to diameter. The crankpin is turned down next, and, after this is done, the inside of the webs are faced off.

The cylinders are turned from 1⅜-inch square cold rolled steel stock and a square flange is left on the base. The cylinder bore should be finished carefully with a hand reamer to insure accuracy and high compression.

The pistons are turned out of 1¼-inch cold rolled steel rod. The stock is turned down to 1⅛ inches and a 1-inch hole bored out to a depth of ¹¹⁄₁₆ inch. After the holes to accommodate the wrist pins are drilled the contact surface of the piston should be carefully polished. ·

The connecting rod is made of two pieces—one of brass and one of steel. The cross piece is cut from brass and drilled with a ¼-inch drill. Only three of the connecting rods are made identical; the fourth being differ-

Fig. 120—The engine with a propeller attached to its shaft

ent at the lower end where it is fixed to the master bearing.

The lower bearings for the three connecting rods are cut from brass stock to a diameter of $^{15}\!/_{32}$ inch and drilled eccentrically with a ¼-inch drill. The bearing on the fourth or master connecting rod is made from a piece of square brass stock with a semicircular cut in one end where it rests on the crankpin after it is fitted into the slot of the master bearing. Both the bearings and cross pieces are secured to the connecting rods by means of silver solder.

The valve casing is turned out of steel to exact diameter with a shoulder at one end to hold the distributor in place. The interior of the opposite end of the valve case is threaded to accommodate a face plate which is drilled in the center with a ¼-inch hole for the steam

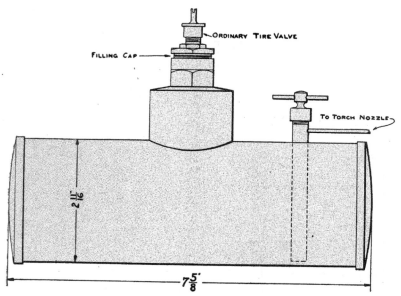

Fig. 121—Drawing of the fuel tank for the gasolene torch

inlet. When the plate is screwed in place, there should be a small space between it and the distributor to act as a steam chest. A small spring is also placed between the plate and the distributor to keep the latter against the shoulder. The steam pipe from the boiler is connected with the face plate by means of a small flange and four screws, interposing a leather gasket. The distributing element should be made to fit the case as perfectly as possible, as a poor fit will cause leakage, which will reduce the efficiency of the engine.

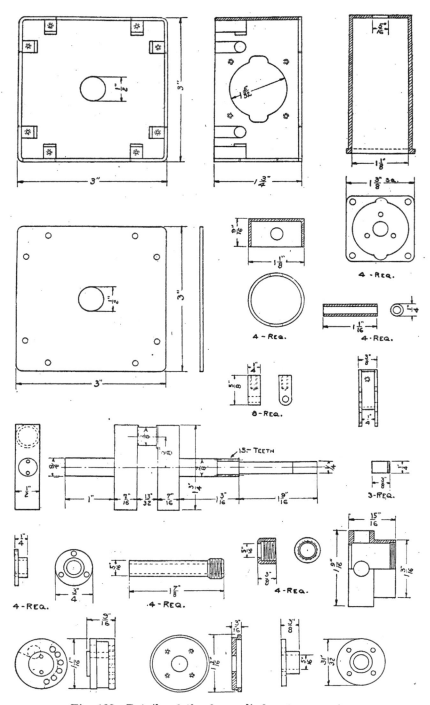

Fig. 122—Details of the four-cylinder steam engine

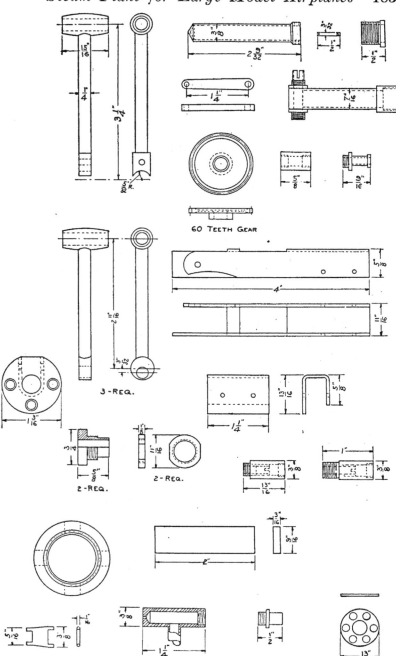

Fig. 123—Details of the four-cylinder steam engine

The distributing tubes are bent by first filling them with either lead, sand or resin—preferably lead for a real good job—the filling being tapped or melted out after bending. One end of each distributing tube is furled to make a steam-tight joint where they are connected to the tubes on the valve case by means of unions.

All parts of the water pump are made of brass with the exception of the frame and clamp which are of sheet steel. The pump proper is held in place by forcing it between the sides of the clamp which is depicted in the detail drawings, this being riveted to the frame. The gear on the pump meshes with the one cut in the crankshaft; the ratio being 4 : 1. The pump discharge is connected to the front end of the inside coil of the boiler.

Having constructed the engine, the next consideration is, of course, the boiler. Where compressed air in large quantities is available, this makes a satisfactory substitute for steam and, except for the cooling effect due to its expansion, is probably superior to steam. But it should be borne in mind that the quantity of air required for a motor of this size is considerable, and it is vital that a heavy supply be available. The same is true regarding carbon dioxide and similar propulsion mediums; they are admirable for testing purposes, but the large consumption of the engine renders steam the most practical operating fluid. It should be mentioned in this connection that in an engine having as high an expansion ratio as this one, the cooling resulting from the employment of gas compressed in cylinders, as in the case with CO_2, becomes serious.

The boiler, shown in Fig. 124, is of the flash type, and is supplied with water from a suitable supply by the engine pump described above. The coil is double and is constructed of $\frac{1}{4}$-inch copper tubing. This is wound on a mandrel, slightly under 1 inch in diameter, to the length of a foot. When completed, the coil will spring open suffi-

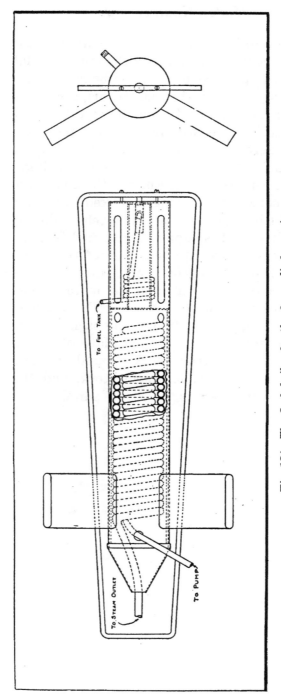

Fig. 124—The flash boiler for the four-cylinder engine

ciently to allow its removal from the mandrel, and will now be of the correct diameter, although some latitude in this respect is permissible. Some heavy gauge sheet brass is wrapped around this coil to bring its diameter up to a scant 2 inches, and the outside coil is wound back over this. The brass is withdrawn after the operation. A casing of galvanized sheet iron 20 inches long is wrapped around the coils as shown in the drawing, which also shows the connections and makes the general assembly clear. A cone 3 inches long is affixed to the front of the casing. The burner is a gasolene torch on the familiar Bunsen principle, and is supplied with fuel from a small tank under air pressure, applied with a tire pump. The preheating and vaporizing coil is of $\frac{1}{4}$-inch seamless steel tubing, given a few turns around the main burner tube of 1-inch steel tubing while red hot. The main tube is about 4 inches long and is fitted at the back with the fuel nozzle and perforated air supply plate shown in the detail drawings. The framework around the casing is of $\frac{1}{4}$-inch square cold rolled steel stock, and the case may be fitted with sockets in which suitable supports can rest.

CHAPTER XV

Operation and design of the engine—How the engine is made—Design for a six-cylinder engine working on the same principle—Description of a flash steam plant to drive the engine.

The development of a suitable power plant for very small model airplanes has long been the ambition of a great many model enthusiasts. To the advanced model maker there is not a more interesting and fascinating branch than that of model airplaning.

The use of steam as a motive power for airplanes is, as we all know, a very old idea, having been first attempted by Stringfellow in 1846 in England, by Langley in 1896 in America, and of late years by Messrs. W. O. Manning, H. H. Groves, and V. E. Johnson in England. The experiments of these last three gentlemen, as recorded in the *Model Engineer and Electrician,* have been a great incentive and also exceedingly helpful and instructive to the writer in the development of the flash steam plant about to be described.

Fig. 126 shows the general arrangement of the engine, boiler, fuel container, and burner.

The engine is of the single-acting rotary type, having three cylinders—the bore ⅝ inch and the stroke ¾ inch. The cylinders are of Tobin bronze, bored out quite thin, and supported radially from the valve by hollow columns on one side, and by the propeller bracket on the other. Needless to say, the valve is of the rotary type which admits steam to each cylinder successively as the pipes, or supporting columns, pass over the inlet port. As shown in Fig. 130, steam is admitted during 50 per cent

187

of the stroke, or while the engine revolves 90 degrees; then the cut-off takes place and the engine works on the expansion of the steam during the next 90-degree turn.

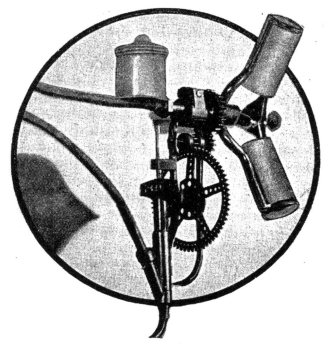

Fig. 125—The three-cylinder engine mounted on a model airplane

The exhaust port is then uncovered during practically the entire return stroke of the piston.

The valve also forms the main and only bearing on which the engine revolves, and is made in a taper form so that any leakage or wear may readily be taken up. The crank is screwed and locked on the end of the valve. This allows the former to be set for either direction of rotation, and for the timing of the steam inlet and cut-off.

All joints and pipe connections are silver soldered, as soft solder would soon melt out under the high temperature at which the engine operates. The propeller mounting consists simply of a star-shaped piece of ⁵⁄₆₄-inch steel brazed to the cylinders, and the propeller is held to this by means of a screw and nut. Three driving pins are set in the plate, which assures a positive drive to the propeller.

The weight of the engine alone is 3¼ ounces. This light weight for an engine capable of developing a quarter horse power can be obtained only in one of the radial

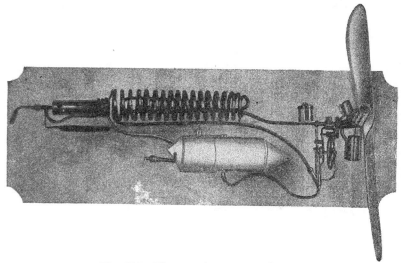

Fig. 126—The complete power plant

cylinder types, because of the absence of a long multi-throw crankshaft with its correspondingly extensive crankcase and bearings.

The flash boiler was made from 8 feet of ⅛-inch inside diameter copper tubing wound up on a tapered mandrel. The over-all length is 7½ inches, being 1⅞ inches in diameter at the burner end and 1⅜ inches at the tail end. The super-heater consists of two loops of the same material placed in the center of the coils, one end of which is brazed to the latter, and the other end fitted with a

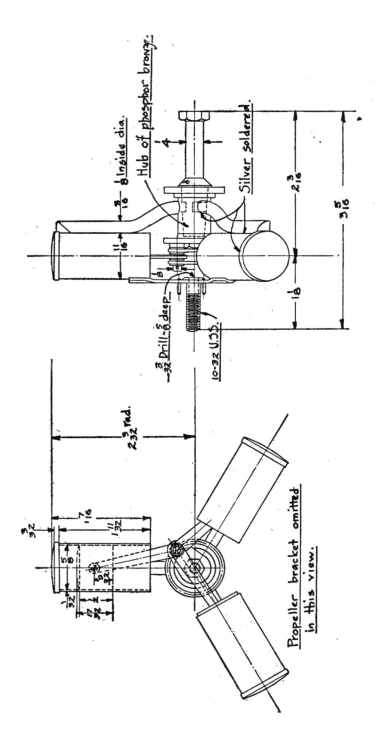

Fig. 127A—Details of the three-cylinder engine

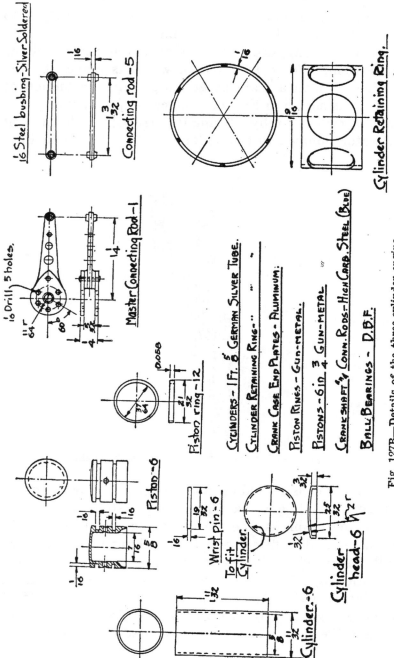

Fig. 127B—Details of the three-cylinder engine

screwed connection for the purpose of connecting it to the hollow crankshaft of the engine.

Water is supplied to the boiler at the big end of the ·coils from a force feed pump, which is driven by the engine through a train of gears, giving a reduction of 5 to 1. As shown in Fig. 128, the crank pin may be set at various radii, thereby altering the stroke; i.e., the amount of water pumped to the boiler. A very rigid and light

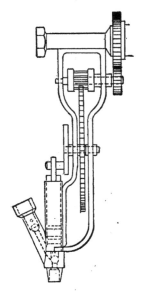

Fig. 128—How the pump is geared to the engine

frame was constructed of umbrella ribbing for the pump support and its gearing. The pump piston and cylinder are made of brass, the former being about ¼ inch in diameter and is fitted with two packing rings. Bronze balls are used in the check valves to prevent corrosion and sticking. A good seating for these valves is obtained by placing a steel ball of the same diameter on the seat and tapping it lightly with a hammer. To limit the movement of the balls, a brass pin is sweated in the side of the valve housing which projects into the water passage at a point slightly above the balls.

The fuel container is made in a stream line form using brass foil .007 of an inch in thickness. There are two compartments, one of which contains 2 ounces of gasolene, the other 4 ounces of water. Allowance is made in the gasolene reservoir for a small air space, so that a slight air pressure might be generated by means of a hand pump in order to start the torch.

The vaporizing coils around the nose of the burner are of light gauge brass tubing. Thin steel tubing is used for the Bunsen tube. A priming tray was beaten

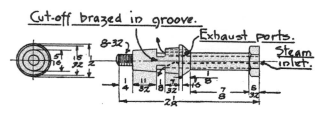

Fig. 129—The engine hub

out of brass foil and supported from the burner by a stiff wire, which was brazed at points indicated in the sketch.

The speed of the engine is largely dependent upon the intensity and amount of heat. To start the plant, the water container is completely filled through the opening at the top. The gasolene container is then filled half full and then given one or two strokes of a hand pump through the check valve which projects from the front of the container.

The needle valve at the burner is then opened long enough to allow the gasolene to drip from the spray nozzle and thence to the priming tray. Half a teaspoonful is sufficient to heat the burner. When this is almost completely burned out, the needle valve may be opened and the burner will start off with a terrific roar. It is then necessary to start the water into the coils before they become overheated. By turning the propeller over several times, the pump is set in operation and the water is

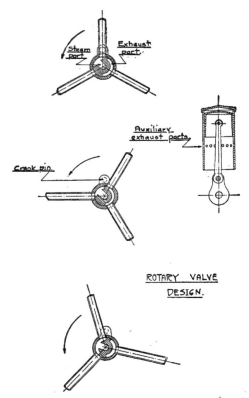

Fig. 130—The rotary valve design for the three-cylinder engine

forced into the hot coils. Considering that the coils work at a red hot temperature, the value of the steam pressure often runs up to between 200 and 300 pounds per square inch.

When the plant is on test, a perforated asbestos casing is placed around the coils and the burner opened wide so as to allow the flame to extend the entire length of

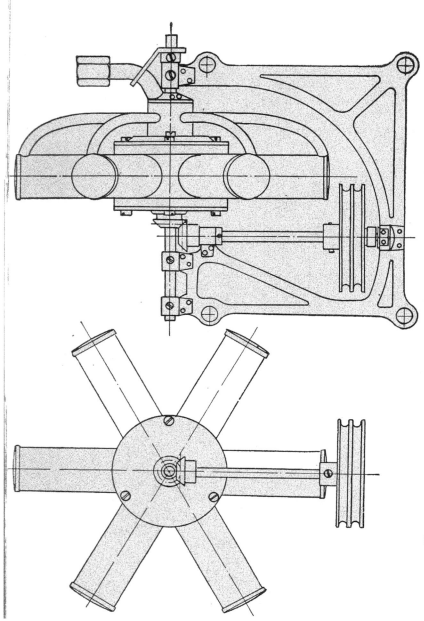

Fig. 131—Design for a six-cylinder engine built on the same principle as the three-cylinder engine

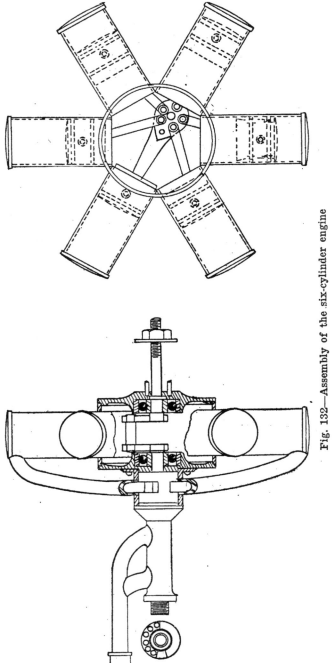

Fig. 132—Assembly of the six-cylinder engine

the coils. After the first flash of steam reaches the engine, it is time to keep clear of the propeller, as it will speed up very rapidly. It would probably be disastrous to the engine to allow it to run light on the flash boiler at full pressure, when 60 pounds per square inch will spin it up to 3,500 R.P.M.

It is very necessary that the engine be lubricated continuously, otherwise the high temperature of the steam would soon cause it to run dry and seize. This is accomplished by a gravity oiler in the steam line shown in Fig. 126, between the engine and boiler. There are two pipe connections to the reservoir, one of which serves to equalize the pressure on the oil and the other as the feeder outlet. A small pointed screw is fitted in the outlet in order to regulate the rate at which the oil drops into the steam line. This adjustment is made through the filling hole at the top of the reservoir.

A MODEL STEAM TURBINE

Description of the design used—Machining the parts—Forming die for turbine buckets.

In designing and constructing a model turbine engine, the model engineer is confronted with several problems which do not have to be taken into consideration when dealing with a reciprocating engine. The most important of these are: The high speed; the balancing of motive parts, adequate bearing and a positive oil supply. In the design here given, these details have been carefully worked out, yet the construction is such that it will not be found difficult of accomplishment by any amateur possessing a small lathe fitted with a plain slide rest.

In the turbine engine the principle that "Action and reaction are equal" is most practically demonstrated. The most efficient turbine engine is the one having a comparatively large number of buckets, but to secure this in a model the size of the one here shown, it would be necessary to turn out the motor wheel with a solid rim, like the flywheel of an engine, and mill out the buckets from the solid metal using an end milling cutter for the purpose. This construction would necessitate the use of special facilities with which few amateurs are equipped.

Drawings A and B, Fig. 134, show the complete model. From these views it will be seen that the steam inlet is cast as a boss on the side of the casing instead of being a separate piece, for which a hole would need to be drilled diagonally through the casing. The exhaust steam passes out at the bottom of the casing, which facilitates the keeping of the interior free from water. The reduction of

198

speed is accomplished by a worm wheel mounted on the turbine shaft and engaging with a gear on the driving shaft. This permits of a reduction of 20 to 1, yet allows an increase of speed if so desired, by substituting a smaller gear on the driving shaft. The side thrust of the

Fig. 133—The complete steam turbine, showing its comparative size

jet of steam on the turbine wheel is offset by the thrust of the worm wheel on the gear of the driving shaft.

The shaft of the turbine wheel is made as long as possible, and the outer end supported in an outboard bearing. Each bearing is fitted with an oil ring and reservoir. Where the shaft passes through the casing, a stuffing box and gland is introduced. As the pressure at this point is only that of the exhaust steam, a piece of felt or candle wicking will be sufficient packing to use.

The construction of the turbine wheel will be considered first. This piece of the model requires to be very

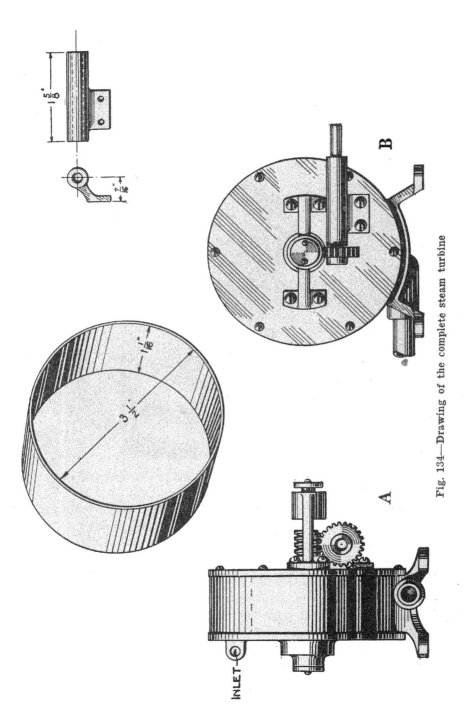

Fig. 134—Drawing of the complete steam turbine

carefully constructed, and when completed, the wheel should be in perfect balance. The design requires forty vanes or buckets arranged around the circumference of the wheel. Fig. 140 shows in detail the construction of the wheel. This wheel is made with a solid web between the hub and rim, instead of being spoked; the reason for this being that the entire surface of the wheel may be turned up true to make it run in perfect balance. The ¼-inch shaft of the model is too small to use in truing up the rough casting of the turbine wheel; it is therefore advis-

Fig. 135—The turbine, showing the exhaust port

able to drill and ream the hub of the casting ⅜ of an inch and use an arbor of that size while machining it.

After the wheel is trued up all over and the rim formed to fit the buckets, a bushing, ⅜ inch outside diameter and with a ¼-inch hole through it, can be made to fit into the hub of the wheel.

This bushing must be carefully made or the wheel will not run true on the shaft. The bushing is shown in Fig. 140. The shape of the bucket is also shown in Fig. 140. The lower portion of the bucket is spherical with a projection which fits into a slot of the wheel. The forming of these buckets will be the greatest difficulty that the amateur will encounter in the construction of the models.

To make the tool for forming the buckets, take two pieces of steel $\frac{3}{8}$ inch x $1\frac{1}{2}$ inches and about 2 inches in length. Clamp the two pieces together and drill two $\frac{1}{4}$-inch holes for guide pins near one end, as shown in Fig. 141. The opposite ends must be squared up and a center punch mark made on one of the pieces $\frac{5}{64}$ inch from the edge where the two pieces join. Drill a small hole to a depth of $\frac{3}{8}$ inch, using a drill about No. 24 size. With the two pieces firmly clamped together, drill into the same hole with a $\frac{1}{2}$-inch drill to the same depth and follow this up with a ball end cutting mill as shown. This can be made from a piece of $\frac{1}{2}$-inch steel and the end rounded to a spherical form. The teeth can be filed in and the tool then hardened and tempered. When using this tool, it will be necessary to remove it from the work often and clear the teeth of the particles of material cut away from the work. If this is not done, the grooves will fill up and the cutter refuse to work. When the depression is cut to the proper depth, a piece of $\frac{1}{2}$-inch steel is turned to the same form as the end of the cutter and inserted in one of the pieces of steel, as shown in Fig. 141. This can be fastened in position by a rivet. This forms a punch and die for forming up the buckets. The material of which the buckets are made should be soft brass of about No. 26 gauge. This should be cut into pieces considerably larger than the size of the buckets, say one inch square. Place the forming tools with the "die" below and lay one of the pieces of soft brass in position on it. Place the "punch" above it on the guide pins, and, holding the three pieces together, transfer them to a larger vise and squeeze together firmly. On relieving the pressure, the sheet of brass will be found in the form shown in Fig. 141, and only requires to have the flat part trimmed away to make it into a bucket of the proper shape for this model.

After the required number have beeen formed up, one

of them should be very carefully laid out and cut to the size and shape shown in Fig. 140, and this can be used as a gauge with which to mark out the others. The cross section of the turbine wheel shows the rim turned out to form the same shape as the bottom of the bucket. This can be somewhat modified as shown where the bucket rests on two projections, one on either side of the rim and the center cut away on a straight line. The latter is

Fig. 136—The turbine, showing the steam inlet

the easier construction and keeps the buckets in position equally well.

After the wheel is turned to the required shape, the slots for the buckets must be made in the rim. If one has a milling machine or a lathe arranged for plain milling, the work is very easy. Most amateurs will be under the necessity of marking off the divisions with dividers and cutting the slots with a hack saw. The divisions should be marked off very carefully and the saw held at right angles to the rim of the wheel when used. After the slots are cut in the wheel and the buckets formed and trimmed, they must be soldered in position. Place the wheel flat on a level surface and set the buckets in place. If the slots have been cut with a hack saw, the thickness of the metal of the bucket will not fill it up. In that case, small pieces of sheet brass can be cut and set in at the front of the bucket. When the buckets are all in place, wrap a piece of iron wire around the outer ends to hold them securely while being soldered. In soldering, a flame can be used direct on the wheel, or it can be done with a tin-smiths' copper. Do not use a large quantity of solder. All that is required is to lock the buckets while their tops are being turned off to nearly the diameter of the inside of the flat brass ring which is to surround them.

Before the buckets are turned off, the wheel should be placed in a permanent position on the shaft. The ends of the shaft should not be turned down to the size of the bearings, however, until the wheel and buckets are entirely finished. The flat ring surrounding the buckets is made from a piece of 3-inch brass tubing about $\frac{1}{16}$ inch in thickness. This should be placed on a block of wood which is held in the jaws of the lathe chuck, or fastened to the face plate, and trued up to a width of $\frac{7}{16}$ inch.

In turning down the tops of the buckets, use a sharp pointed tool and take very light cuts. Turn them down until the flat ring can almost be forced on, then lay the wheel down on a flat surface, heat the ring evenly all around, and it will expand sufficiently to drop over the tops of the buckets. If this is carefully done, no soldering of the buckets to the ring will be necessary. The shaft and wheel are again mounted in the lathe and the wheel carefully turned all over, using a very pointed tool and taking very light cuts. This should put the wheel in perfect balance. Lastly, turn the ends of the shaft to the dimensions shown. When the shaft and wheel are complete and supported lightly between centers, they should stand in any position in which they are placed. Should the wheel revolve, it would indicate that one side of the wheel is heavier than the opposite side. In this case, it will be necessary to balance it by drilling a few small holes in the rim of the heavier side.

The casing of the turbine consists of a piece of 3½-

inch brass tubing about $\frac{1}{16}$ inch thick, shown in Fig. 134. Into each end of this tubing a head is fitted. The tubing should be forced on a wood mandrel turned to tightly fit its inside diameter. Both ends should be trued up parallel, leaving the length of the tubing $^{11}\!\!/_{16}$ inch. To the lower side of this tube is attached the base. The upper part of the base is turned out, or carefully filed, to fit the outer curve of the tubing. A boss is cast on one end of the base, and is drilled out to fit the exhaust pipe. Where the drill comes through, an elongated hole will be formed on the top curve of the base. A corresponding hole should be marked out and cut through the side of the tubing, and after the four holes in the feet are drilled, the two pieces are soldered together.

The two heads for the casing cannot be made from the same pattern without a considerable waste of material and extra work. For the head on the "steam" side of the turbine J, the pattern should be made with a straight hub for a chuck piece. This is for the purpose of holding the casting firmly in the chuck while it is turned to fit the tubing; the inside surface turned off; the oil reservoir finished, and the $\frac{1}{4}$-inch hole for the bearing bored through. All these operations must be done at one setting; i.e., the castings must not be disturbed in the chuck until all these operations are finished. This is absolutely necessary in order that the parts shall all "line up" when the work is completed.

The $\frac{1}{4}$-inch hole for the bearing should not be made by starting a $\frac{1}{4}$-inch drill and cutting the hole while the work revolves, for the hole would in all probability run crooked. Start with a smaller drill, say $\frac{3}{16}$ inch, and when that has been put through, use a small boring tool in the slide rest of the lathe and true up the hole, taking repeated cuts until it is almost to the required size, then run a $\frac{1}{4}$-inch reamer through to finish it. The outside diameter of the head should be left a trifle larger than the 3½-inch tubing, as it will add to the appearance of the finished model.

When cutting out the oil reservoirs, turn out the groove for the brass washer shown in Fig. 140. This washer is to be soldered in place, but this must not be done until the oiling ring for the bearing, L, Fig. 139, has been placed in the aperture and the screw holes of the heads drilled as will be explained later. After finishing the inside of the head, it can be reversed in the chuck and held by the inside surface of the flange while the

Fig. 137—The turbine taken apart, showing the construction of the rotor

extra part of the chuck piece is cut away and the end squared up. Two holes are to be drilled into the oil reservoir, one above, for the purpose of supplying oil, and the one below to be used to drain the reservoir. They should be drilled with a No. 42 twist drill and tapped with a No. 4-36 tap. Machine screws can be used to stop these holes when the model is in operation.

The bearing sleeve is clearly shown at L, and a cross section of the same is seen at J. For this purpose, a piece of brass rod can be used. Two pieces are to be made, one for each bearing. In addition to the length required for the bearing, the pieces should be cut off of sufficient length to hold in the chuck. When the extra length is gripped in the chuck jaw, center the end of the revolving piece with a sharp pointed tool and start a smaller drill than the size of hole required, after which the hole is trued up with a very small boring tool held in the slide rest and finished with a reamer. Next turn down the outside to $\frac{1}{4}$ inch and square up the shoulder of the outer end. The bearing is held in place by the two small adjusting screws as shown in J and L. The groove for the oil ring can be cut with a file or hack saw, and the reamer must again be inserted to remove the burr. The oil ring can be made from a piece of tubing.

The next important operation on this head is to drill and ream the steam nozzle in the lug cast on the outer surface. The line of the center of this nozzle should be at an angle of twenty degrees with the inner surface of the head as shown at M and N. M is a horizontal section cut through the center of the lug. This shows the manner in which the hole is drilled. The position of the hole is laid out on the inside of the head as shown at J. It should be on a direct vertical line with, and $^{15}/_{16}$ inch above the center of the head. Mark the point with a center punch. When this has been done, a small piece of metal, shown at N, should be soldered on just outside. This is to prevent the drill from running off to one side when it is started. A center punch mark should be made on the outer end of the lug at the proper point to bring the finished hole at the proper angle with the turbine wheel. Against the center punch mark the point of the back center of the lathe is placed and the hole drilled. When drilling the hole, the head must be carefully held up or its weight will break the drill. Use a No. 60 twist drill. The hole is next reamed out carefully with a taper reamer from the inside of the head. The hole should be $\frac{1}{16}$ inch in diameter at that portion where it emerges into the casing. The outer end can be counterbored from the outside to fit the steam pipe, and the entrance to the nozzle from the pipe beveled off as shown.

The other head requires less work. All the turning can be done at one setting of the casting in the chuck. The projection of the stuffing box on the inside of the head is to be used as a chuck piece. This is firmly grasped in the jaws of the chuck. The outer diameter of the head is first turned down to the same size as the opposite head. The inside flange is next turned down to the same diameter as the inside of the 3½-inch tubing. This can be done by using a bent tool in the tool post of the slide rest. It can be gauged for size by using a pair of calipers and setting them to fit the finishing diameter of the flange of the other head. The gauging of the piece of work by the calipers must be done while it stands at rest, for if done when the work is revolving, the tendency will be to leave the flange too large, as the calipers pass over a revolving piece of work with greater ease than if stationary.

The outside surface of the head is next turned off true and the recess for the gland bored out to the proper size. The small hole for the shaft should next be put through and this should be done very carefully, as this will aid materially in the final lining up of the parts when the model is assembled, as will be described later. The hole should first be drilled with a smaller drill than the finished size called for, after which a small boring tool is used to true up the hole and fit it to the exact diameter of the shaft of the turbine wheel. The reason for doing this is that when the outboard bearing of the shaft is completed and ready to attach to the casing, the wheel and shaft can be inserted, and this hole in the stuffing box will hold the shaft in alignment until the bearing can be set into place and the position of the holes for the screws marked on the head. When the outboard bearing has been fitted, the hole in the stuffing box should be enlarged a little so that the shaft will not touch it.

The packing gland is shown at Q and will not require an extended explanation. It should be a sliding fit in the stuffing box and the hole should be at least $\frac{1}{32}$ inch larger than the diameter of the shaft. Two clearance holes for No. 2-56 screws should be drilled in the flange and tapped holes to correspond marked off and drilled

Fig. 138—The castings for the model steam turbine. The pressed buckets are also shown

in the head. No finishing is required on the inside of this head.

The two heads can now be marked off and drilled for the screws which are to hold the casing together. Lay off six points around the edge of the flange of the head first described J, and mark them with a center punch. Drill these holes through, using a No. 42 twist drill. When all six holes are drilled, place the flanges of the two heads together and hold them in position with a clamp. The projection of the stuffing box on one head will enter the oil reservoirs turned in the other head. With the heads securely held together, use one of the holes as a guide for the drill and start a hole in the other head. This must not go deeper than the bevel of the lip of the drill. Mark the other five holes in the head in the same manner, after which the two heads can be separated.

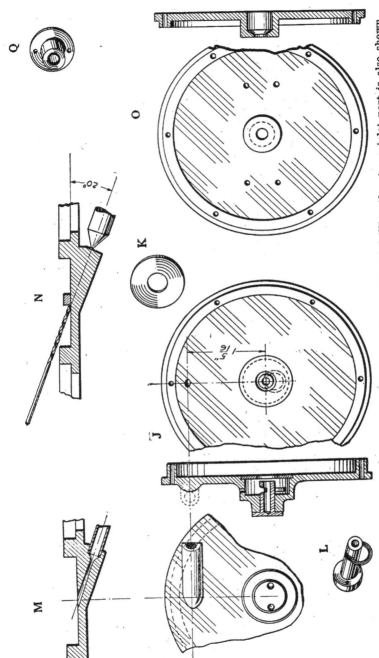

Fig. 139—How the steam side of the turbine is finished. The method of drilling the steam inlet port is also shown

The holes marked should be drilled through with a No. 32 twist drill, this being a clearance hole for a No. 4 screw. The holes drilled with the No. 42 drill are next tapped with a No. 4-36 tap.

After these holes are drilled, the washer can be soldered in place to form the oil reservoir in the head J. The casing can then be assembled with the wheel and shaft in place.

The outboard bearing should be held in the jaws of the chuck by the outer end of the hub so that the feet project toward the slide rest. Rotate the lathe head by hand and measure with a tool in the slide rest to see that the piece is chucked so that the feet project equally. With a centering tool in the slide rest, mark a center in the revolving hub of the casting, and, using the same size drill as used for the hole at J, put a hole entirely through to size it from one of the bearings shown at L.

The oil reservoir is next bored out and the little depression made for the washer. The washer, like the corresponding one in the other bearing, must not be soldered into place until the oil ring has been placed inside. The bottom of the feet must be faced off while the casting is in the chuck, light cuts with a pointed tool being made so as not to loosen the work. The piece can then be removed from the chuck and the bearing sleeve fitted. The holes for two adjusting screws are next drilled and tapped, after which the clearance holes for the screws are placed in the feet. The bearing can now be slipped onto the end of the shaft projecting from the assembled casings and the position of the screw holes transferred to the casing.

The head O must now be removed and the holes drilled and tapped. The hole in the back of the stuffing box can be enlarged, as mentioned, and the head replaced. The worm wheel should next be placed on the shaft in the

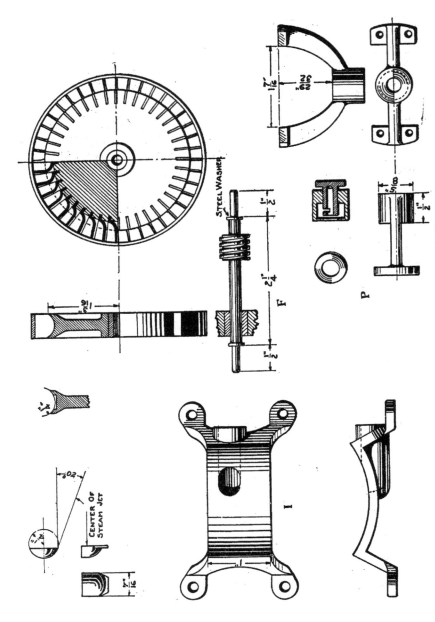

Fig. 140—Details of the steam turbine

position shown in Fig. 140 and secured by a very small set screw drilled and tapped into one end of the worm.

The bearing for the driving shaft is best finished by centering at both ends and drilling the hole when the casting is held against the back center of the lathe. The drill used should be a trifle smaller than ¼ inch. Be careful that the drill does not touch the end of the back center. When the hole is nearly through, a piece of hard wood or metal should be interposed to prevent injury to the drill or center. The hole can then be reamed with a

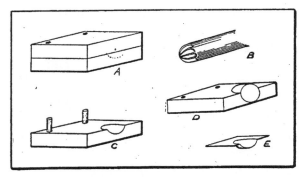

Fig. 141—Forming device for the brass buckets used on the rotor of the steam turbine

¼-inch reamer, after which it can be placed on a ¼-inch mandrel and the ends squared up. If desired, the outside of the bearing can be turned down and finished on either end. If finish is desired over that portion where the flange is attached it will be necessary to do this with a file, as it cannot be turned. The back of the flange would be filed up flat and parallel to the hole for the shaft, after which two screw holes are drilled.

The gear wheel and shaft are fastened together by a pin or screw and placed in position in the bearing. These parts are then held in place on the casing until the positions of the screw holes are located. The points of the teeth of the gear should not be allowed to "bottom" on the worm, yet they should have sufficient contact.

If desired, a small sheet metal oil pan can be formed to fit under the gear wheel for lubrication and a cover could be made to enclose all the gearing, but there are so many who prefer to "see the wheels go 'round" that these parts have not been shown. No driving wheel is shown on the end of the driving shaft. This must be made of the proper size to give the required speed to whatever model or piece of apparatus it is required to drive.

When the final adjustment of the parts is made, the turbine wheel should run just as close as possible to the head containing the nozzle, but without touching it. Perhaps the easier way to make this adjustment is to loosen the screws in the bearing sleeve of the "steam" side and screw up the adjusting screws in the outboard bearing until the turbine wheel rubs against the side of the casing when the shaft is revolved by hand. Then turn these screws back a half revolution and carefully tighten up the screws on the "steam" side until the bearing sleeve comes against the shoulder of the shaft. Be careful that the sleeves are not forced up too tight, as it is preferable to have a little play endwise on the shaft, though this should not exceed ⅟₆₄ inch. A stop cock should be placed on the steam pipe as near the model as convenient.

CHAPTER XVII

DESIGN AND CONSTRUCTION OF MODEL BOILERS

Efficiency of model boilers—Evaporative power—Convection currents—
Boiler design—Pot boilers—Water-tube boilers—Marine boilers—
Riveting model boilers—Super-heaters.

Small boilers to drive model steam engines may be made according to several different designs. Some of these are more efficient than others from the standpoint of evaporative power and some are more adaptable than others for a specific purpose. As an example, a vertical pot boiler would not be as successful for use in a model

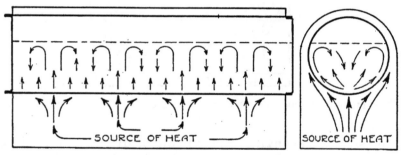

Fig. 142A—Showing the path of convection currents in a pot boiler

speed boat as a flash boiler would be and yet, many times, the flash boiler could not be employed as conveniently as the pot boiler in certain work. The model maker should therefore choose the type of boiler that is most adaptable to the work for which he wishes to use it. The following paragraphs will describe a few of the more common types of boilers together with information that will aid greatly in designing and constructing them.

First, a few words in regard to the efficiency of boilers

in general. Boiler efficiency depends upon the evaporative power and this, in turn, depends upon several things, such as the heating surface, the source of heat and the design of the boiler. The evaporative power of any boiler, whether model or large, is measured by the number of cubic inches of water that it changes into steam during a period of one minute, and from this, calculations may be made as to the power of the boiler and

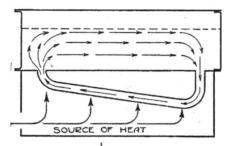

Fig. 142B—Showing the path of convection currents in a water-tube boiler

the size of the engine it is capable of driving. Model pot boilers (about 7 inches long by 3 inches in diameter) cannot evaporate more than ⅓ to ⅜ of a cubic inch of water per minute. This figures out about 1 cubic inch of water per every 60 square inches of heating surface. This efficiency is considered very low when compared with a tube boiler fired with a blow lamp. Such a boiler will consume or evaporate as much as 1 square inch of water per 30 square inches of heating surface. The low efficiency of an ordinary boiler is due to the fact that it has a comparatively small heating surface in ratio to its size. Water tube boilers are much more efficient.

When water is heated in a vessel, violent currents are set up within the water and such currents are called "convection currents." These currents tend to move from the surface to which the heat is applied. Their movement in a pot boiler is shown in Fig. 142A.

The motion and rapidity of these convection currents is a large factor when the efficiency of a boiler is considered. The faster these currents move, the better the circulation of the water will be, and this is to be desired. The boiler shown in Fig. 142B is the type known as the water-tube boiler and has a greater efficiency than the ordinary pot boiler because convection currents are set up in its tubes as shown and these currents permit a maximum transmission of heat from the fire or flame to

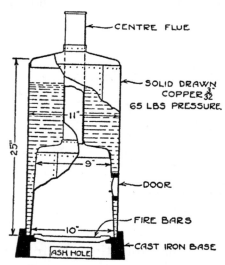

Fig. 143—A single-flue boiler made of drawn copper tube

the water. For this reason, boilers embodying this principle have a much greater efficiency and evaporate power than those of ordinary design.

A boiler known as the single-flue type is shown in Fig. 143. This resembles a pot boiler and, in fact, its efficiency is not much greater. The construction of this boiler, however, if followed carefully, will give the model maker much valuable knowledge in practical boiler construction, as it embodies all the good features of a well-made boiler from the standpoint of design.

The boiler proper may be made of either brass, copper or steel. Of the three metals, copper is by far the most suitable for model boiler construction. In boilers of small size and for low pressures, the seams and joints may be silver soldered, but in the larger types it is necessary to rivet the parts together. When rivets are employed, they should have a diameter of at least 1¼ times the thickness of the boiler plate, no matter what metal is used in its construction. This procedure will give the proper factor of safety. The rivets should not be placed

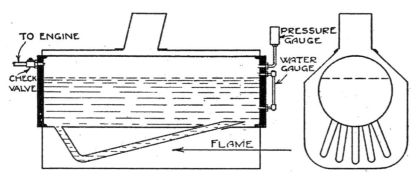

Fig. 144—A simple water-tube marine boiler

any farther apart than four times their diameter. When boiler joints are silver soldered, they are generally riveted first to hold the seam in place until the solder is applied. For small boilers, silver soldering is to be recommended, as it gives a steam-tight seam and will also withstand considerable pressure without rupture. In many cases, when a very small boiler is to be made, the model maker can employ a solid drawn copper tube, which may be obtained on the open market. Such tubes are very strong and will stand up under great pressure. Brass and steel tubes of this nature can also be used.

When metal tubing is used for the boiler, considerable care will have to be exercised in putting in the end pieces of the boiler, as these are more apt to succumb to the steam pressure than the tubing itself. There are several methods of putting the end pieces in. In boilers with low steam pressure, a metal disc can be cut which will fit into the end of the tube with a driving fit and, after it is in place, it can be silver soldered. In no case can ordinary solder be employed, as when this substance

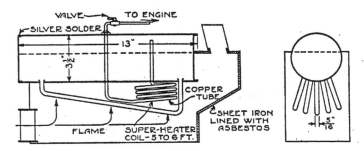

Fig. 145—A water-tube boiler equipped with a super-heater coil

is heated appreciably its tensile strength is reduced greatly and it is therefore unable to withstand any great pressure. If the ends of the boilers are put in place in this manner, it is very good practice to fix a brass rod concentrically in the boiler, leaving the ends projecting and providing them with a nut which will bear part of the pressure which is exerted on the end pieces. In many cases, the seams and joints of the boiler may be welded or brazed, but as this necessitates the use of an elaborate equipment the average model maker is not able to follow this practice. A flanged end plate riveted to the boiler is much more effective than the ordinary disc and this procedure is advisable when boilers of a larger type are constructed. A good substitute for a flanged end plate is shown in Fig. 150. The ring or strip is riveted to the boiler tube and the end plate rests against this.

Now that the general procedure in boiler construction has been outlined, a brief description of several suitable types of model steam engines will be given. The one shown in Fig. 155 is of the marine type and in actual practice is found to steam very well, although its evaporating power is not as great as that of a flash boiler. The boiler shown is of the semi-flash type and being that it sets very low it is especially adapted to model steamers as it would help keep the center of gravity low, and this

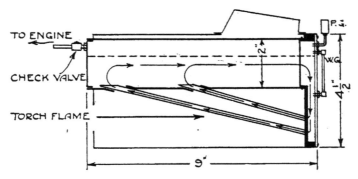

Fig. 146—A water-tube boiler with a cast waterback

adds to the stability of any craft. It will be noticed that the fire box is in the shape of a tube which runs through the boiler. Running through the fire box, at right angles to each other, are six water tubes upon which the flame of the blow torch impinges. Water circulates through these boiler coils continuously when heat is applied to them. The fire tube should be made of steel as this metal is less affected by heat, whereas copper and brass oxide much more rapidly. The water tubes are placed in the large steel tube by means of silver solder. To make a more substantial job, the water tubes should be cut a little long and when they are put in place the ends may be flanged as shown in the detail drawing of the boiler. One end of the steel fire box is provided with a funnel which is riveted in place. A small hood also extends out over the back of the boiler, which protects the flame from external air currents. The boiler should be provided with a safety valve and pressure gauge. A water glass should also be fitted to the back and the valve placed on the top to check the steam flow to the engine.

Fig. 146 shows a very suitable type, although it presents more difficulties in actual construction than the one previously described. It is necessary to have a casting made for the waterback ring, which is provided with a flange to which the boiler tube is riveted. The waterback ring is provided with six holes drilled as shown in the drawing. The water tubes are threaded and screwed into these holes and run from this point to the bottom

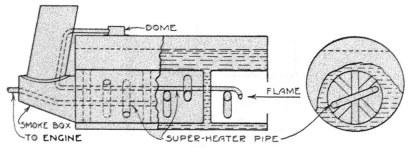

Fig. 147—A marine boiler provided with a super-heater loop

of the boiler. The method of fastening them in the bottom of the boiler is shown very clearly in the figure. A hole the size of the tube is first drilled in the bottom of the boiler and then a steel rod the same size as the tube is inserted in this hole and the rod is then bent to the same position that the water tubes will be in. After the holes are treated in this manner, the water tubes are inserted in them and they are silver soldered into place. This particular part of the construction is very simple, as the builder will not have to bend the water tubes into any particular shape. The boiler proper can be made of solid drawn copper tubing with an internal diameter of about 2 inches. The copper boiler tubes should have a thickness of about ⅛ inch and this will give a very high factor of safety. The back plate of the waterback ring can be cut from steel and it should have a thickness of at least ⁵⁄₁₆ inch to prevent it from bulging when the steam pressure reaches its maximum value. The back

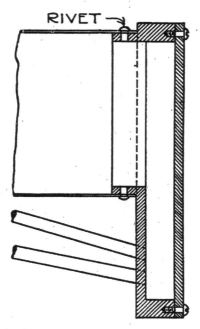

RIVET

Fig. 148—A waterback for a water-tube boiler, showing how the boiler tube is riveted in place

plate is screwed to the waterback casting. The whole boiler is enclosed in a casing bent into shape from Russian iron and a small funnel is riveted to the top of the casing to carry away the fumes and gases that are generated by the torch or fire. The inside of this casing of Russian iron should be heat insulated by means of sheet asbestos, which is fixed to its inner surface.

The feed pipe for the engine is attached to the free end of the boiler and this pipe should be provided with a valve. A water glass and a steam pressure gauge should be affixed to the back plate of the boiler. The circulation of water in this particular type of boiler is very good and for the ordinary model steamer, where great speed is not sought, this boiler is recommended.

The boiler shown in Fig. 145 is a conventional type and differs from the others described in that it is pro-

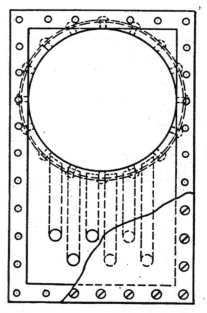

Fig. 149—Another drawing of the waterback shown in Fig. 148

vided with a super-heater coil. The boiler proper can be constructed in the ordinary manner and it is provided with five water tubes bent and inserted as shown in the drawing of the device. Being that these water tubes are the same shape it is well to make a small wooden form to bend them on so that they will all be uniform. It is also advisable to put these tubes in place in the bottom of the boiler before the end pieces are soldered in and after the super-heater coil has been wound and soldered into position in the bottom of the boiler. It will then be possible to silver solder the joint where the tubes pass through the boiler on both the inside and outside, insuring a good, steam-tight joint. The tubes are arranged, as shown in the rear view of the boiler. The tubes can be of copper or steel and they should be about ⁵⁄₁₆ inch in diameter for use with a boiler of the dimensions shown.

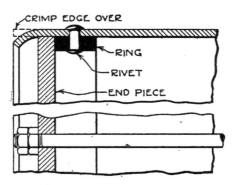

Fig. 150—A very practical method of fastening boiler end plates in place

The fire box of the boiler is made of Russian iron lined with sheet asbestos and provided with a funnel at the forward end.

A modification of this particular type of boiler is shown in Fig. 144. The water tubes in this are arranged in the same manner as those on the boiler shown in Fig. 145. This boiler, however, is not provided with a super-heater coil and is therefore not as efficient as the one previously mentioned. The entire boiler is enclosed in a casing made from Russian iron and lined with sheet asbestos to keep it insulated. The casing of this boiler is not square, but of the shape shown. A boiler casing made according to this outline will not have as much metal in it as one made square and, therefore, some saving in weight is effected.

The boiler shown in Fig. 147 is very similar to that illustrated in Fig. 155, with the exception that it is provided with a small super-heater loop which runs from the dome on the top through the water tubes in the fire box and leaves the front of the casing to the engine. The

Fig. 151—Casting of a waterback

tubes in this boiler are arranged just a little different than those in the boiler shown in Fig. 155. Three tubes are arranged vertically while the remaining five are arranged at 45 degrees and cross each other at right angles. With this exception, the other features of this boiler can be constructed like that shown in Fig. 155.

The boiler in Fig. 153 is a modified Scott type, which is a sort of combination flash and tube boiler. The coils or, rather, the loops of tubing beneath the drum, make the boiler a very quick steamer and, while its water is exhausted rather quickly, a pump added to it makes the outfit a presentable one. As it stands, the boiler holds sufficient water for a satisfying run in a race or demonstration. The reader will understand that this description is supposed to pertain to the construction of the boiler, rather than to be a eulogy of its fine points.

The drum is a length of brass tubing $2\frac{1}{4}$ inches in diameter and $8\frac{1}{8}$ inches long. The holes for the under loops or tubes are best drilled by packing the brass tube with a wooden mandrel which will assist in center punching and drilling, after the position of each hole has been marked.

The loops are of standard copper automobile tubing of the $\frac{1}{4}$-inch size. This material is so delightfully soft

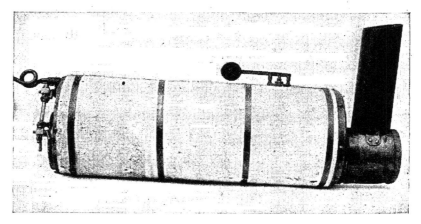

Fig. 152—A complete marine boiler of the type shown in Fig. 147

and easy to work that no particular trouble will be experienced in bending it to the required loop-shape. The reader will notice that the loops are in the form of horseshoe magnets with one leg shorter than the other. The loops are placed with the long and short arms alternated to provide for a natural thermo-syphon circulation of the hot water through the cold.

To bend the loops, a jig or form should be constructed of hard wood by sawing out the profile with a jig saw. This will insure uniformity, which is at least desirable if not actually essential. The loops are preferably silver

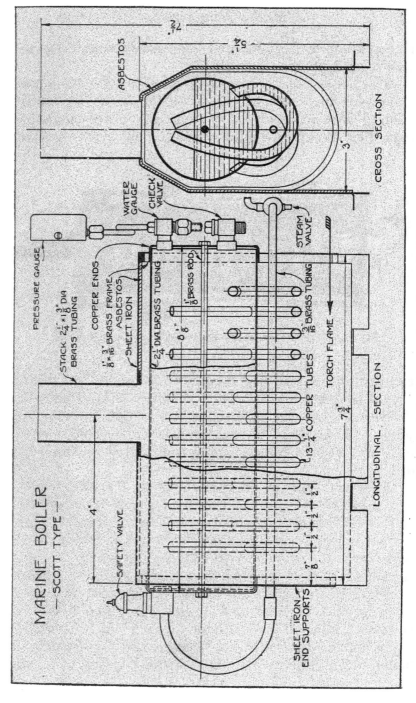

Fig. 153—A Scott-type marine boiler, with fittings

soldered into the boiler drum. This is essential if a gasolene torch is to supply the heat.

The boiler heads are of copper discs with the edges spun over to a tight fit on the drum. They are then secured with a long stud running the entire length of the boiler. Finally the joint is made doubly secure against

Fig. 154—The bottom of the Scott boiler shown in Fig. 153

rupture and leakage by careful silver soldering. Care should be taken to make sure that the joint is actually "sweated," i.e., that the solder flows right through the joint to the inside of the boiler. The proper heat, a good flux, and a clean piece of work will certify to this condition.

From one head of the boiler is taken the safety valve and super-heater pipe. This super-heater is necessary, as the steam from this boiler is necessarily wet and unfit for use unless some means may be found to dry it. The pipe solves the problem as it conducts the steam through the hottest part of the fire before passing it to the engine.

Flash steam boilers are entirely different from any of those previously described in that no actual boiler is employed, as a coil of small pipe through which the water is circulated and heated replaces the more conventional boiler form. The average flash boiler consists of a coil of either steel or copper tubing from 2 to 3 inches in diameter containing from 8 to 30 feet of tubing. The amount of tubing in a flash boiler depends entirely upon the power that it is required to deliver. With a small engine with a ⅝-inch bore, 10 feet of ⁵⁄₁₆-inch tubing would be sufficient to drive it at maximum power. It is calculated that 30 feet of tubing the same diameter and furnished with enough heat is sufficient to generate approximately 1 H.P. Working on the figures given above, the model maker should be able to construct and design a flash boiler for any small engine, either single or double cylinder. Another factor to be considered, however, is the type of

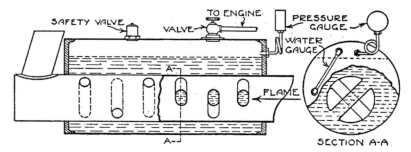

Fig. 155—A marine boiler with fittings

engine, whether it is single or double acting, as this has much to do with the steam consumption.

The boiler coils for a flash steam plant should be made of steel, as this does not oxidize as rapidly as copper tubing and will therefore last much longer, although there is no serious objection to the use of copper other than this disadvantage. The steel tubing may be wound on a wooden form with a diameter of the inside dimensions that the tube is to be. The wooden form is held in the vise and the tube is bent around it. Before this is done, however, the tube should be put through a process of annealing so that it will not crack or break while being wound. The tube may also be wound over a mandrel on the lathe, the lathe being turned with one hand slowly and the tubing guided on the mandrel with the other. One end of the boiler coil should be provided with a check valve and the other end is provided with a coupling, by means of which it is connected to the water supply pipe.

CHAPTER XVIII

MODEL BOILER FITTINGS

Design and construction of safety valves, check valves, water cocks, water gauges and steam gauges.

The efficiency of a model boiler depends somewhat upon its fittings, such as safety valves, stop valves, check valves, et cetera. A poorly constructed, leaky valve or boiler fitting of any kind is just as bad as a leak in the boiler and will cause a serious reduction in the working pressure. Therefore, it is quite necessary that a valve or fitting of any kind be made very accurately to insure maximum efficiency.

Probably the most important fitting on a model boiler is the safety valve. There are many types of valves, each with its advantages and its disadvantages. In some cases, one type of valve is more adaptable for a certain boiler than another. A properly designed and regulated valve will blow off when the critical pressure is reached in the boiler; below this pressure it should not leak or "weep" in the least. After the valve functions, it should return to its normal position immediately the pressure has fallen beyond the danger-point. Such a valve is not very easy to construct, although the model maker, at first thought, may think that it does not present any difficulties whatsoever. In designing safety valves the model maker must bear in mind the fact that the interior surface of the valve plug determines the pressure at which it will blo woff—the larger the surface, the lower the pressure needed, and vice versa.

A very simple type of valve is shown in Fig. 156, at A. Although this is not strongly recommended for use on

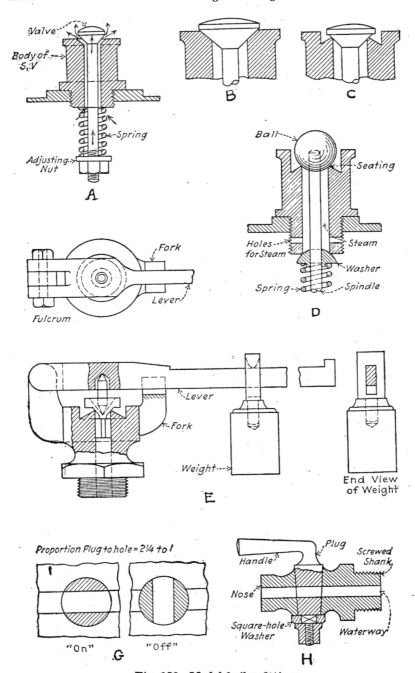

Fig. 156—Model boiler fittings

a large boiler, it will be found suitable on the smaller models working at very low pressure. The working pressure of the valve is regulated by means of the spring shown. This is done by moving the nut either up or down; down if the pressure is to be reduced and up if the pressure is to be increased. The spring should not be made of steel, as this will corrode and rust by the action of the steam, and it will be found that brass spring wire will be the only suitable metal for this purpose. The body of the valve should screw tightly into the boiler to prevent leakage, and great care should be taken in making the seating accurate. The main objection to a valve of this type is the large area of contact in the seating. The larger the area of contact, the more difficult it will be to make the seating accurate and steam-tight. The least speck of dirt or projection on the metallic surface will cause the valve to leak badly.

A much better valve seating is shown at G. This is generally known as the "knife-edge" seating and makes a more leak-proof valve than that described in the foregoing paragraph. The contacting surface is very much smaller and therefore the possibility of the valve becoming fouled with dirt is much more remote. Another valve on this principle is shown at D. This is known as the "ball-valve" and is very efficient for model boilers. The seat is made extremely sharp. The ball is of bronze and may be purchased on the open market. It is attached to a spindle, to the opposite end of which is fixed the usual form of spring. To make this seating as accurate as possible, a steel ball the same size as the bronze ball is placed over the valve orifice and given a sharp blow with a hammer. This slightly concaves the surface of the valve seat and makes a much better fit possible, thereby increasing the efficiency of the valve. The body of the valve may be turned from brass stock.

Valves employing springs are not suitable for model

boilers of the larger type, as they are not dependable enough. Then there is the disadvantage of having the adjusting spring within the boiler. Larger boilers, for this reason, are generally provided with a weight valve. Such a valve, together with its various parts, is shown at E. The "blowing off" point of this particular valve can be easily regulated by either moving the weight on the lever closer or farther away from the fulcrum. As the weight is moved away, the pressure necessary to "blow" the valve will be greater, and vice versa. Such valves cannot be used on model boat boilers unless some method is employed to keep the weight in the position that it is set in. By carefully considering the drawing, the reader will find no trouble in setting about to make such a valve, as it involves no complications, but the workmanship must be good. Safety valves, as a whole, are cranky things to fuss with, and they cannot be made too accurately. They must be absolutely steam-tight and at the same time they must not stick.

A very good marine safety valve of the spring type is shown at F, in Fig. 157. This valve, unlike many others of the spring type, possesses the advantage of being adjustable from the outside by means of the nipple shown. A discharge pipe for the exhausted steam is attached to the side of the valve casing and leads up the side of the ship's funnel if the valve is used on a marine boiler. This exhaust pipe should be plenty large enough to effect a free and easy discharge of the steam. The proportion shown in the drawing is about right. So much for safety valves.

There are innumerable designs for model steam and water cocks, and it frequently behooves the model engineer just what type is most suitable for the problem in hand. The two sketches shown at G, in Fig. 156, illustrate the "on" and "off" position of a simple one-way valve. There is one thing that the model engineer should

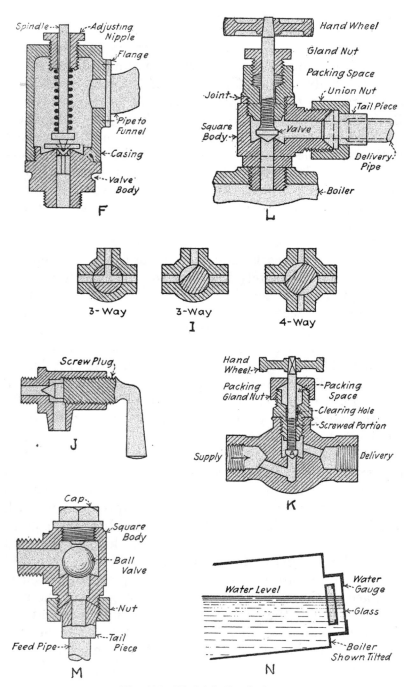

Fig. 157—Model boiler fittings

always remember in connection with the making of cocks; the moving member should always be of a different metal than the body. Thus if the body of the cock is made of gunmetal the plug should be made of brass or bronze. Bronze and good brass rod also work very well together. Another point demanding the consideration of the worker is that the hole in the plug should not be out of proportion to the plug itself. The proportion shown at G is about right and if this is departed from it would be better to make the opening or hole in the plug smaller rather than larger.

A simple, straight-nosed plug cock is shown at H. The body of the cock should be made first and then the plug can be accurately fitted in so that it will provide sufficient lap to effect complete stoppage of the water, gas, steam or whatever may be passing through it. The method employed in producing the tapered hole may not present itself to the reader at first thought, and a few words will be said in regard to it. The hole is first drilled with an ordinary drill of the proper size and then it is reamed out with a small tapered reamer made especially for the work. The conical reamer for such a job can be cut from a piece of steel rod and hardened. The end of the rod that has the taper is filed half round to provide a cutting edge. Several such reamers can be made up for cocks of different sizes. The plug is then turned to shape and properly drilled in a small improvised jig with a drill the same size as that used to produce the hole in the body of the cock. The bottom of the plug is filled square for a short distance and below this it is threaded to receive a small nut, which holds a square-hole washer in place on the square portion of the plug above the nut. The purpose of the square-hole washer is to prevent the nut from tightening when the plug is rotated. Three-way and four-way valves are often employed in model engineering, and such types are shown at I, Fig. 157. These valves involve practically the same problems as those met with in constructing the simple one-way cock.

A very simple one-way cock of an entirely different type than that previously described is shown at J, Fig. 157. This is much easier to construct, as will be readily seen by a glance at the drawing. The body may be made of gunmetal and the screw-plug can be turned from brass. If the threads are cut accurately on this plug, it is much less liable to leak than plugs made by the other method. The workman will understand that the plug is turned from a piece of rod and threaded, after which the handle is bent over. It may be necessary to employ a small bending jig to do this. Valves of this type are especially recommended for high-pressure boilers.

In some cases, valves of the type described above are not suitable for use, and the more conventional "globe" or "screw-down" valve must be employed. Such a valve is shown at K, Fig. 157. Owing to the small diameter of the passages it will be impossible to core them out during the casting of the body and it will therefore be necessary to drill them. An especially made tool will have to be used in making the valve seat. The "needle" is filed square at the outer end to receive a small hand wheel with a knurled edge. A valve very similar to this but with the passages differently arranged is shown at L. This is easier to make, as the holes are not so difficult to drill.

A check valve is a valve that will freely permit water to pass in one direction, but entirely prevents its passage in the opposite direction, the latter claim being made with the assumption that the valve is designed and constructed properly. A check valve of the ball type is shown at M. The body can be cast and drilled out. A special tool will also have to be used here to form the valve seating. The valve proper is a small brass ball of the correct size. The normal position of this ball is shown, but when water enters from the feed-pipe the ball is forced up against the surface of the small cap screwed in the top of the valve body. If, however, water attempts to pass in the opposite direction due to back pressure, the ball immediately returns to its seating and thereby closes the passage. The course to pursue in constructing this valve is so obvious that a detailed description of the procedure is considered unnecessary.

There is probably no other model boiler fitting that offers more problems to the model engineer than the water glass or gauge. On very small water glasses, great trouble is experienced due to capillary action of the water in the glass tube, as this sometimes entirely overcomes the gravitational force that would otherwise keep the water at the proper level—that which corresponds with the level of the water inside the boiler. The operation of a water gauge is shown at N, in Fig. 157. A simple water gauge is illustrated at O, in Fig. 158, and the principal part of this is a glass tube of small bore bent as shown. The bending of the glass tube can be done in a hot flame, and if the builder is not sufficiently acquainted with the bending of glass he may call upon some of his chemical friends. Unless done by an experienced person, the glass is very apt to kink at the corners. This water gauge is not suitable for very large boilers. The one shown at *P* is a much better design and does not present any more difficulties in construction. The packing gland should be put in place very carefully to prevent leakage, as this is very apt to occur if the steam pressure is high. Unless the parts are very accurately made, and the holes in both the upper and lower portion are in perfect alignment, the glass tube is apt to be broken when it is put in place. The joints of this device are silver soldered, as noted in the sketch. A very elaborate water gauge appears at Q. This is especially suitable for boilers of the larger type. It is provided with one-way cocks at each end, as well as cleaning plugs, which are characteristic of the larger devices. The cocks are provided to check the water should the glass become broken through accident. The cock at the bottom is used to clear the glass of mud or dirt if the water level becomes obscure through this cause. It is very advisable to fit water gauges with small guards not only to prevent the glass from being broken on the outside, but to offer some protection to the operator should the steam pressure become so high that the glass is unable to hold it. This is not a remote possibility, and owing to the very fragile nature of glass, the pieces would fly in every direction.

A very simple form of hand force pump for feeding a boiler is illustrated at R, in Fig. 158. A valuable feature of this pump is the means provided to limit its stroke. When used in connection with a steam pressure of over 50 pounds the bore of this pump should not be over ¾ inch, as otherwise it will be most difficult to operate by hand owing to the steam pressure that will have to be overcome. The delivery valve is provided with a ¹⁄₁₆-inch projection across the top to obviate the possibility of its blocking up the passage to the union when it rises. A small recess is filed at the bottom of the pump to admit the water.

A more powerful hand pump for use with higher steam pressures is shown at S. This force pump also has a greater pumping capacity than the one previously described. The plunger of this pump is very long and is provided with several packing glands and double cup leathers. By altering the length of the operating lever, this pump can be used on quite high pressures, as the longer the handle is the greater the leverage and thereby the greater force the operator is able to overcome.

A simple form of pressure gauge for model boilers appears at U. This operates on the "Bourdon" principle, i.e., a steam pressure inside a bent tube tends to

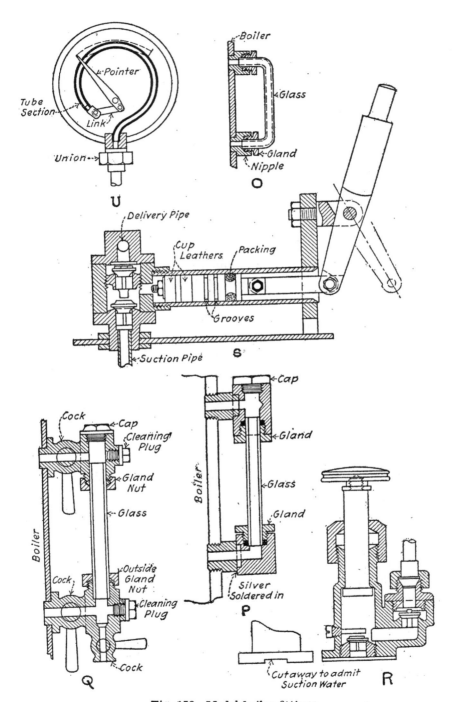

Fig. 158—Model boiler fittings

straighten the tube out. Advantage is taken of this by
arranging the elements of the pressure gauge as shown.
The higher the steam pressure is in the tube, the more
its tendency will be to straighten out. This causes the
pointer which is connected with the tube to move across
a dial which is properly calibrated to indicate the steam
pressure in pounds per square inch. The writer has heard
of instances where an ordinary automobile tire gauge
was employed successfully to indicate steam pressure on
a model boiler.

CHAPTER XIX

A RECORD-BREAKING MODEL HYDROPLANE

The hull of the boat—Its flash boiler and twin-cylinder, high-speed
steam engine.

The model described in this chapter is the result of considerable experimentation, both in the making of various types of hulls and power plant equipment. The craft was really made to bring the model speed boat record from England to America, and in tests "Elmara" has

Fig. 159—"Elmara" at a thirty-mile clip

shown a speed slightly in excess of 30 miles per hour. The English record held by the "Evil Spirit" is 26.7 miles per hour.

The dimensions of the hull are as follows: Length 39.37 inches, beam 7⅞ inches, step 1¼ inches high, sides forward of the step 4¼ inches high, sides directly back of the step, 3 inches high. The distance from the bow to the step is 17¼ inches. The weight of the complete hull is 2 pounds 1½ ounces.

Only two materials are used in the construction of the hull—aluminum and mahogany. Mahogany is a very strong wood, will take a smooth finish and is more or less

Fig. 160—Showing the power plant of ''Elmara''

unaffected by moisture. The pattern of the sides of the boat are first cut out of paper and this paper is pasted on a piece of dressed mahogany ⅛ inch thick. The mahogany is then cut into shape. The bow piece is cut out of solid mahogany and shaped as shown in Fig. 164.

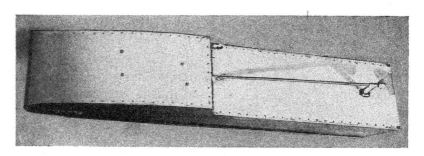

Fig. 161—The bottom of the boat

Square mahogany strips are then cut out and fastened to the inside of the side piece by means of shellac and ⅜-inch brass brads. The bottom of the craft is made of No. 22 gauge sheet aluminum and this is fastened to the square mahogany strips, as the sides of the boat are only ⅛ inch thick and it would be next to impossible to fasten the aluminum to these without splitting them. The aluminum is also fastened by means of shellac and ⅜-inch brass brads. The shellac tends to make the boat watertight, while the brads hold the aluminum rigidly in place. The aluminum bottom does not run completely over the bow piece, but merely overlaps it sufficiently to be fastened to it by means of the brass brads. The single step in the bottom of the boat is formed by a mahogany strip through which the propeller shaft tube runs and the water scoop. The back of the boat is also made up of

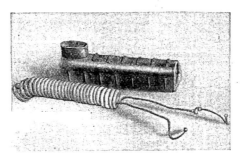

Fig. 162—The boiler coil and boiler casing of the power plant

mahogany. A small aluminum hood is bent into shape and this is fixed to the bow of the boat and prevents water from reaching the engine and also reduces air resistance.

The builder is cautioned to use extreme care in making this hull, as every detail must be paid attention to in the construction of model racing boats, and a hull put together carelessly cannot be expected to attain great speed. Attention must be paid to the most minute details, such as perfect balance, wind resistance, water resistance, etc. The wooden portion of the hull should be rubbed down well and thoroughly shellacked, applying the shellac with a camel's-hair brush so that it will leave the surface bright

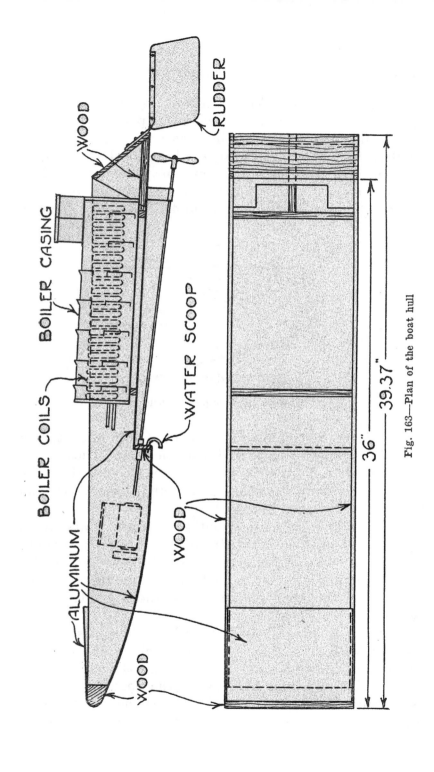

RUDDER

WOOD

BOILER CASING

BOILER COILS

WATER SCOOP

ALUMINUM

WOOD

WOOD

WOOD

36"

39.37"

Fig. 163—Plan of the boat hull

and smooth. The hull of "Elmara" is so designed that it will plane at a speed of fifteen miles per hour.

While some experimenting has been done with propellers for use on this boat, the best results have been obtained with a cast aluminum propeller of 3⅝₆-inch diameter and a pitch of 10.2. When the craft is at full speed, the propeller turns over at 4,000 R.P.M. The pro-

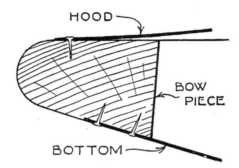

Fig. 164—How the bow piece is held in place

peller shaft or stern tube is made of ⅜-inch brass tubing with brass bushings soldered in at each end. These bushings are drilled for a ³⁄₁₆-inch silvered steel shaft. The skeg is made up from flat and round brass stock and screwed to the skeg bearer.

The rudder can be made either of brass or aluminum and is fixed in place so that the boat will run in a straight course. It will be found necessary to bend the rudder slightly to one side, as the boat has a noticeable tendency to turn in the same direction that the propeller is revolving.

The flash steam boiler of "Elmara" consists of 18 feet of ⁵⁄₁₆-inch O.D. seamless steel tubing of 22 gauge. This is wound in a single spiral on a 1¾-inch mandrel. To avoid trouble it is advisable to thoroughly anneal the steel tubing before it is wound. The casing, which covers the boiler coil, is bent into shape from 22-gauge Russian

iron. The ends are all lap-seamed and the funnel is
flanged at the bottom and riveted to the boiler casing.
In making the casing, a clearance of at least ⅜ inch
should be left all around the boiler coil. When completed,
the casing is lined with sheet asbestos having a thickness
of ⅛ inch. The boiler is mounted on a small frame made
of square mahogany strips.

A little trouble will be had in making the boat keep
a straight course when in operation and the experience
of the builder has shown that if the propeller is mounted
just a little off center in the opposite direction to which
the propeller turns this will have a tendency to make the

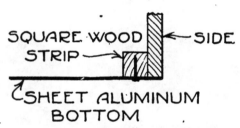

Fig. 165—Showing how the aluminum bottom is held to
the sides of the hull

boat follow a straight course. The amount of offset
necessary will have to be proven by experiment, and it
will be found that very little is necessary.

If the builder desires, the bow piece can be made considerably lighter by boring a large hole through it with
an auger. In fastening the bottom to this piece, the wood
should be cut away so that the surface of the aluminum
will be flush with the surface of the bow piece. Otherwise
the edge of the aluminum will have a tendency to prevent
the boat from planing, owing to the resistance it would
give to the water.

The water scoop consists of a small semicircular
piece of copper tubing with an internal diameter of ¼
inch and arranged in the wooden portion of the step.

The water scoop is connected to a small water tank, which will be described later.

Unless the builder is anxious to keep the weight of the hull down as low as possible, it is advisable to cover the sides next to the boiler with asbestos or light sheet aluminum to prevent it from catching fire from the heat of the blow torch. It is very inconvenient to have this happen

Fig. 166—The twin-cylinder engine used on the boat

when the craft is in the center of the lake and no row-boat is handy.

The power plant of the boat is especially interesting from the standpoint of model steam engineering. It possesses several unique and original features of construction and operation, although a glance at the photograph may convince the reader that it is nothing extraordinary.

The main castings of the engine are of aluminum. Only two castings are used in the engine itself—the upper or cylinder portion and the crankcase. The cylinders of the engine, which have a bore of $1\frac{1}{16}$ inch, are cut from Shelby steel tubing. The cylinders are reamed out and lapped before they are inserted in the casting, as shown in the cross-section drawing of the engine, Fig. 171. The main casting is, of course, carefully bored or drilled out

Fig. 167—The oil and water tank for the engine and boiler

so that a good driving fit is made. The lower portion of the steel tubing is turned to a slight taper and this holds the cylinder rigidly in place. The casting is so made that after the cylinders are in place there will be a recess around them, and this is made to act as an exhaust chamber, as will be explained later in connection with the valve action. The cylinders are provided with 20 auxiliary exhaust ports which are drilled with a $\frac{1}{16}$-inch drill and spaced equadistant. At the extreme limit of the down-stroke, the cylinders uncover these auxiliary exhaust ports and this places them in communication with the exhaust chamber, permitting about 60 per cent of the steam left in the cylinders to leave. The drawing shows the one cylinder in the exhaust position with the auxiliary ports completely uncovered. If the cylinders project any at the top of the casting after they are driven in place, they should be ground off perfectly flush with the top of the engine, as the cover plate, which is described later, must lay absolutely flat to prevent a possible leakage. A small recess $\frac{1}{16}$-inch deep is filed in the side of each cylinder át the top and the cylinders are so mounted in the casting that these recesses will be exactly opposite each other so that they will form a communication with the steam chest, as will be explained more fully in connection with the valve mechanism.

The crankshaft is turned from a solid piece of cold roll steel. The connecting rods are also turned into shape from cold roll steel. A split brass bearing is used on the crankshaft end of the connecting rods and these are held together by means of two small machine screws, one on each side. The lower portion of the brass bearing is counter-bored to receive the round head of the machine screws used. The screws extend through both halves of the bearing into the end of the connecting rod which is drilled and tapped to receive them. While this method may hold the bearing in place, the original engine has small pins driven through the upper half of the bearing into the lower end of the connecting rod. The pistons are turned to size from cold roll steel and bored out. This gives a solid piston which is absolutely necessary with an engine of this nature working on high-pressure flash steam. A previous power plant of "Elmara" had an engine with the tops of the pistons silver soldered into place. One day the engine refused to go and investigation showed that one of the piston tops had come off, the extreme heat of the flash steam having melted it or come so close to melting it that its tensile strength fell below the critical point. It is best to make a solid piston in the first place as flash steam is unsatisfactory with silver solder. The pistons are provided with piston rings, and these are an absolute necessity with the high pressure employed. Each piston has two piston rings. A hole is drilled completely through the top of the pistons to accommodate the wrist pin, which is a small piece of cold roll steel rod of the proper size. The ends of the wrist pin are filed to conform with the internal outline of the cylinder and finished off smoothly so that they will not scratch the cylinder walls.

The small brass bevel gear which drives the valve shaft is drilled and pinned to the crankshaft. Very little strain in actual operation makes it unnecessary to key

Fig. 168—The engine, showing the hand water pump and the oil pump

this member to the crankshaft. A small universal joint is placed on the crankshaft just back of the bevel gear and this is also pinned into place with as heavy a pin as the quarter-inch shaft will allow. Owing to the fact that this universal joint forms the connection with the propeller shaft, it has to hold up under considerable force, as nearly a horse-power of energy is transmitted to the propeller when the boat is at high speed. This method of fixing the universal joint to the crankshaft is open to criticism, but in the case of the "Elmara" the trouble experienced with it has been exceedingly small in comparison with the trouble had from other sources. The universal joint used is made by the Boston Gear Works and is the smallest stock joint made by these people.

A very special method was employed to hold the flywheel on the crankshaft. The spur gear on the end of the shaft which meshes with the larger pump gear is soldered to a sleeve and the flywheel is drilled out and reamed so that this sleeve will pass into the hole in the flywheel with a driving fit. The sleeve is then drilled out and reamed so that it will fit on the crankshaft with a

Fig. 169—End view of the engine

driving fit. These members are then mounted upon the shaft and a hole drilled through the hub of the flywheel, the sleeve and the crankshaft. A small steel pin is then driven in this hole, which holds the flywheel in place, as well as the spur gear.

The valve will now be explained, together with the steam chest. The steam chest is formed by a small box cut out from solid steel and provided with a cover plate of steel which projects over the sides. The overlapping portion of the plate is drilled so that it may be fastened to the top of the engine by means of the long machine

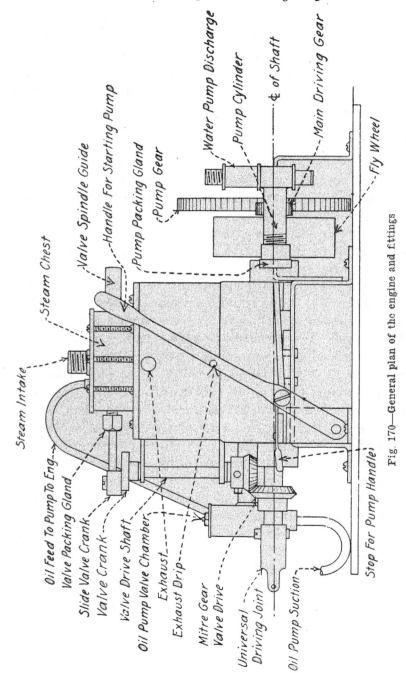

Water Pump Discharge

Pump Cylinder

℄ of Shaft

Main Driving Gear

Fly Wheel

Handle For Starting Pump

Valve Spindle Guide

Pump Packing Gland

Pump Gear

Steam Chest

Steam Intake

Oil Feed To PumpTo Eng.
Valve Packing Gland
Slide Valve Crank
Valve Crank
Valve Drive Shaft
Oil Pump Valve Chamber
Exhaust
Exhaust Drip
Mitre Gear
Valve Drive
Universal
Driving Joint
Oil Pump Suction

Stop For Pump Handle

Fig. 170—General plan of the engine and fittings

screws, as clearly shown in the drawing of the engine. The amount of the projection should not be made too great, otherwise the cover plate will have a tendency to buckle up in the center when the machine screws are tightened. A hole is drilled in each end of the stream chest, one for the valve spindle and the other for the packing gland. The valve spindle guide is turned from brass. All the remaining parts of the steam chest and valve are made from steel. The valve is extremely simple and consists merely of a small block of steel with a recess chiseled in the center as shown. The adjusting nut on the valve spindle rests between two shoulders on the valve so that the stroke of the valve or, rather, its oscillating motion, can be regulated. After the threads on the valve spindle are cut, it will be necessary to turn those on the end of the spindle off that fits into the spindle guide. The valve spindle extends through the packing gland and has attached to its outer end a small slot cut from steel. The eccentric or valve crank consists merely of a small steel disc mounted on the upper end of the valve shaft with a small machine screw placed between its periphery and center. This screw passes through a tiny steel block, which fits in the slot on the valve spindle. The small steel disc has a collar by means of which it is pinned to the valve shaft. This collar rests on a small projection on the end of the engine casting which forms a bearing for the valve shaft. A similar projection forms the lower bearing. These bearings are not bushed, as this has been found unnecessary. At the lower end of the valve shaft, another small brass bevel gear is fixed with a steel pin, and this meshes with a similar brass gear that was placed on the crankshaft previously.

Three slots are cut in the top plate of the engine, one large one with two smaller ones on each side. The small ones are the inlet ports from the steam chest, and the larger one is the exhaust port. The two small slots form

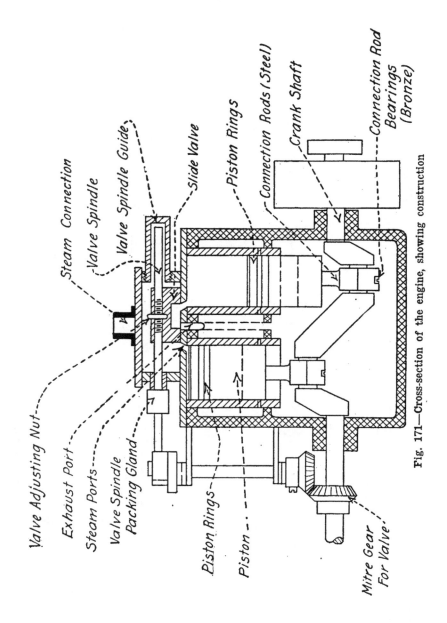

Valve Adjusting Nut

Steam Connection

Valve Spindle

Valve Spindle Guide

Slide Valve

Piston Rings

Connection Rods (Steel)

Crank Shaft

Connection Rod Bearings (Bronze)

Exhaust Port

Steam Ports

Valve Spindle Packing Gland

Piston Rings

Piston

Mitre Gear For Valve

Fig. 171—Cross-section of the engine, showing construction

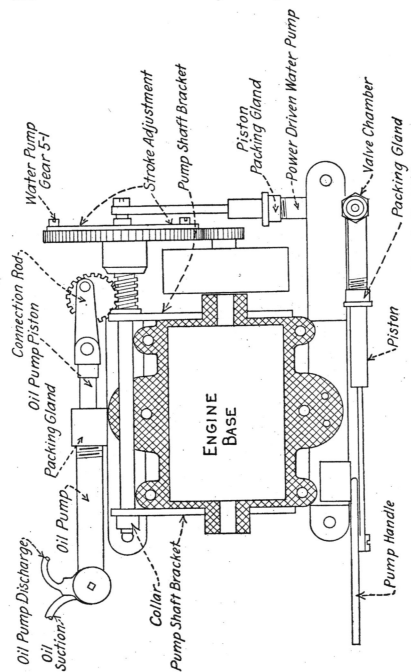

Fig. 172—How the oil and water pumps are driven by the engine

Water Pump
Gear 5-1

Stroke Adjustment

Pump Shaft Bracket

Piston
Packing Gland

Power Driven Water Pump

Valve Chamber

Packing Gland

Connection Rod

Oil Pump Piston

Packing Gland

Oil Pump

Oil Pump Discharge

Oil Suction

Collar

Pump Shaft Bracket

ENGINE BASE

Piston

Pump Handle

a passage to the cylinders by means of the two recesses filed in the top of the cylinders, as explained in a previous paragraph. These slots come directly over the recesses, as shown in the cross section drawing of the engine, Fig. 171. Between the pistons, and directly under the larger exhaust port, a ¼-inch hole is drilled. The exhaust slot communicates with this hole and the engines

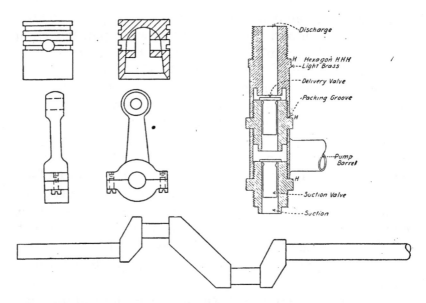

Fig. 173—The crankshaft, connecting rods, pistons and water pump of the engine

exhaust steam therefore comes out at both sides. Between the cylinders and on each side of the engine casting, two ¼-inch holes are drilled from beneath up through the casting until they meet the horizontal hole that was previously drilled between the cylinders. This enables the exhaust steam from the auxiliary ports to pass into the regular exhaust of the engine. After these two holes are drilled, the lower ends are stopped with a brass plug.

Another small hole is drilled in the side of the engine casting, which communicates with the exhaust chamber and this acts as a drip for the steam that condenses into water. By using this method, lagging on the cylinders has been found unnecessary and the efficiency of the engine is not impaired in the least.

An auxiliary structure of sheet metal holds the pumps and pumping mechanism to the engine. Two water pumps are provided, one being the hand operated pump to start the engine with and the other the power operated pump driving off the main shaft through a train of two gears. The small spur gear fastened to the shaft in front of the flywheel drives the larger gear to which the connecting rod of the water pump is attached. The side arrangement for altering the stroke of the pump is arranged for. An arm is attached to the gear, being pivoted to one end with a screw having a slot in the opposite end through which the machine screw is passed. In the center of the arm, the connecting rod of the pump is attached. By using a set screw on the arm and bringing it either closer or farther away from the center of the gear wheel adjusts the stroke of the pump within quite a wide range. The hand water pump is actuated by a large steel lever or handle.

The larger gear wheel which drives the water pump, is fixed on the end of the shaft, which has one of its bearings at the opposite end of the engine. Mounted directly behind the larger gear wheel on the steel shaft is a worm gear which meshes with a small spur gear used to drive the pump. The connecting rod of the pump is attached to the edge of this spur gear, the ratio being 100 : 1. A small pipe leads through the oil pump directly into the steam chest of the engine, where the oil mixes itself with the hot steam and is carried into the cylinders of the engine. The gears of the power water pump have a ratio of 5 : 1. The bore of the water pump is ¼ inch.

The stroke of the water pump is variable from ⅛ to 9/16 inch and the bore is ¼ inch. The oil pump also has a bore of ¼ inch and a stroke of ⅛ inch to ¼ inch.

The complete engine is mounted on an aluminum plate with a thickness of ⅛ inch and this is anchored on cross pieces or bearers in the boat hull. Four machine screws, one on each corner of the base, is used to hold it.

In the design and construction of this engine, the builder is more or less indebted to Mr. H. H. Groves and Mr. Westmoreland, of England, who are pioneers in the use of flash steam and whose instructive articles have appeared in past issues of the *Model Engineer*.

CHAPTER XX

Building the boat hull—The power plant and construction of the deck fittings.

The model described in this article is that of a bulk freighter of canal size such as used in the transportation of grain and ore on the Great Lakes, particularly between Fort William and Montreal. The overall length of the prototype is 260 feet and larger boats of this nature are made up to 650 feet in length.

The hull of the model is 4 feet long overall and the length between the perpendiculars is 3 feet 9¾ inches. The beam at the water line is 8 inches and the draught extreme is 4½ inches. The displacement at this draught is 40¼ pounds in fresh water. It will be necessary to use some ballast on the model in order to bring her down to the designed water line.

In the following lines only a few general constructional hints will be given, as the drawings are complete in every detail and, with these as an aid, the constructor will not have much trouble in making the vessel. The table of offsets which is given herewith should be referred to constantly, and with this as a key to the complete plan shown in Fig. 174, the builder will be able to proceed intelligently.

The hull can be very easily produced by the bread-and-butter principle. For those who are not familiar with this method of construction, a few words will be advisable. The hull is made up of ten planks and each plank is gouged out with a chisel. The plank which forms the bottom of the boat is not gouged, but is shaped with

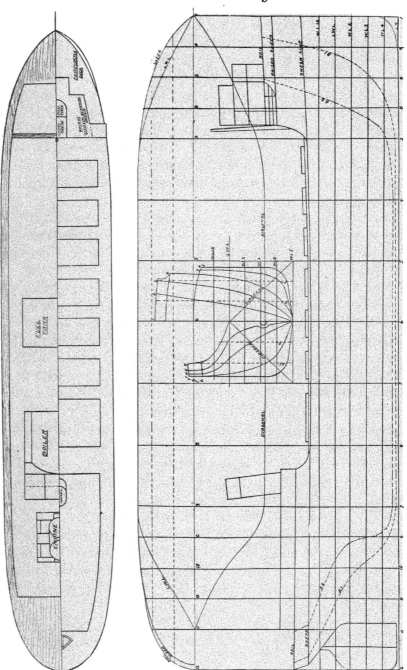

Fig. 174—The plan of the lake freighter, showing water and sheer lines

a draw-knife to conform with the outline shown in Fig. 176. The next plank is gouged out and this is then glued to the bottom plank and so on until the entire hull is built up. After the shapes are drawn on the $7/8$-inch pine planks used, the planks are roughed up nearly to shape with a draw-knife and after they are all produced in this manner their surfaces are smeared with glue and put together with small brass brads. The brads are placed 1 inch apart. The hull is then finished with a plane and sandpaper, being brought to a smooth surface before the paint is applied.

The deck is made from a piece of $1/4$-inch pine board and the hatch openings are cut in this. It will be noticed that there are seven hatches. Six of these are provided for loading the hull and the seventh one, toward the stern of the boat, is much larger than the rest and is intended for making adjustments on the power plant. The deck is held to shape and in place on the hull by the deck beams which are mortised into the side of the hull. A rub-streak of $1/4$-inch square pine is tacked on each side just below the sheer. The sides of the hatches and covers can be made from cigar box wood and may be held in place by means of glue and small tacks.

The deck house, chart house and wheel house, as well as the bridge, are made of tin, bent and soldered into shape. The bridge is equipped with spray cloths made of fine white linen. The port lights in the deck house and sides of the hull are made of brass and provided with pieces of mica glued in place to represent glass.

The life boats carried on top of the engine casing are whittled out of a solid piece of wood and painted white and properly lettered and numbered. The life boats are held by means of string to brackets bent into shape from iron stove-pipe wire.

The engine is of the two-cylinder marine type and has a 1-inch more and a 1-inch stroke and drives a

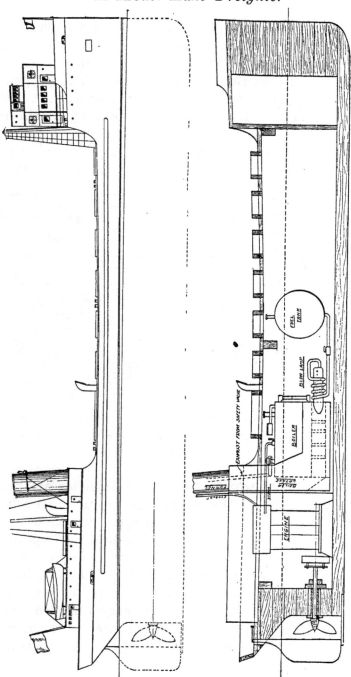

Fig. 174A—General plan of the vessel, showing arrangement of power plant.

4-inch propeller 3 inches in diameter, with a pitch of 2¾ inches. This is fastened to the propeller shaft by means of a lock nut. The propeller will be a matter of experiment until the best results are obtained. The propeller shaft is a ¼-inch steel rod and the stuffing box is turned from solid brass with a cavity for packing at the inboard end.

The boiler is made from seamless copper tubing and is 4 inches in diameter by 5½ inches long. It is provided

TABLE OF OFFSETS																	
STATIONS		1	A	2	B	3	4	5	6	7	8	9	C	10	D	11	12
HEIGHTS	RAIL	9 3/4	9 5/8	9 1/2								7 1/4	7 5/16	7 3/8	7 7/16	7 7/16	7 1/2
	RAISED SHEER	9	8 7/8	8 3/4	8 5/8												
	SHEER LINE	6 7/8	6 3/4	6 11/16	6 9/16	6 1/2	6 5/16	6 1/4	6 3/16	6 3/16	6 1/4	6 5/16	6 5/16	6 3/8	6 3/8	6 3/8	6 1/2
	W.L.1.A.	5 5/8	—	—	—	—	—	—	—	—	—	—	—	—	—	—	5 5/8
	L.W.L.	4 1/2	—	—	—	—	—	—	—	—	—	—	—	—	—	—	4 1/2
	B.OF K.	0	—	—	—	—	—	—	—	—	—	—	—	—	—	—	0
	B 1		6 3/8	2	3/4	5/16	3/16	—	—	—	—	3/16	5/8	2 7/8	4 7/16	5 1/16	
	B 2			9 9/16	3 7/16	1	1/2	—	—	—	—	1/2	1 7/8	4 7/8	5 9/16	7 1/16	
HALF BREADTHS	RAIL		2 1/16	3								4	4	3 3/4	3 1/2	3	
	RAISED SHEER		1 7/8	2 15/16	3 1/2												
	SHEER LINE		1 5/8	2 11/16	3 1/2	3 3/4	4	—	—	—	—	4	4	3 13/16	3 1/2	2 7/8	
	W.L.1A.		1 7/16	2 1/2	3 5/16	3 3/4	4	—	—	—	—	4	3 7/8	3 1/2	3	2 1/4	
	L.W.L.	0	1 1/4	2 1/4	3 1/4	3 3/4	4	—	—	—	—	4	3 3/4	2 5/8	1 1/4	0	
	W.L 2		1	2	3	3 5/8	4	—	—	—	—	4	3 1/2	1 3/4	5/8	0	
	WL 3		3/4	1 5/8	2 5/8	3 1/2	4	—	—	—	—	4	3 1/8	1 1/8	3/8		
	WL 4		3/8	1 1/8	1 7/8	3 1/8	3 7/8	—	—	—	—	3 7/8	2 3/8	1/2	1/8		
	WL 5																
STATIONS SPACED 4 1/2 APART. WATER LINES SPACED 1 1/8 APART.																	

Fig. 175—The table of offsets which should be constantly referred to in building the boat

with one fire tube which is 1 inch in diameter and four cross tubes ¼ inch in diameter. All the tubes are silver soldered into place. The boiler is fired by a blow torch using gasolene under pressure. The boiler is equipped with a safety valve, filling plug, pressure gauge and water gauge. The funnel measures 1 inch wide by 1½ inches long, and extends 4 inches above the engine casing. The exhaust steam from the engine is carried up the waste pipe aft of the funnel. The deck fittings forward consist of a steering boom, two bullards, two fairheads and four life buoys for the bridge. On the main deck are six bullards and two cowl ventilators ½ inch in diameter. The fore mast is properly stayed and fitted with rat-lines. The main mast is properly fitted and stayed and the rigging consists of silk fishing line. The rudder of

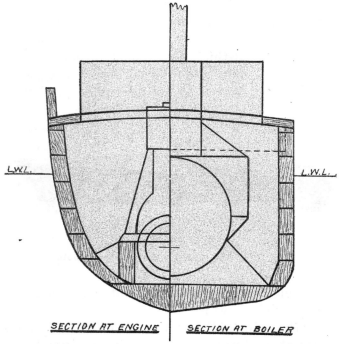

Fig. 176—Sections of the boat at the boiler and the engine

the boat is made of sheet brass and is fitted with a quadrant.

The hull of the model is painted black above the water line and red below the water line. The deck and hatches are painted deep maroon and the chart house, wheel house and engine casing are painted black. The funnel is painted red with a black top. The ventilators are painted white on the outside and red on the inside.

CHAPTER XXI

A SHARPIE-TYPE MODEL BOAT

Making the mahogany hull — Power plant — Construction of special alcohol burner.

The majority of model boat builders lack the skill to get a block of wood, shape the outside and then get down to the real job of hollowing it out with gouging tools to the desired thickness without breaking through. What is wanted for a model speed boat is a type of hull capable of being driven at considerable speed with the

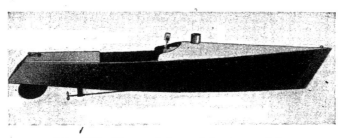

Fig. 177—The model speed boat complete

least possible water resistance and consequently with the least possible power.

The "Experiment" has no curves except the sides. Floor and sheer lines are absolutely straight from bow to stern. A glance at the drawing will give one the impression that the boat is utterly unsuitable for speed, yet the boat has run at the rate of 6½ miles an hour with an ordinary boiler and engine.

To begin with, make full-size drawings of the boat; its plan and cross section. Make two wooden molds, one for the cross section about one-third of the distance from

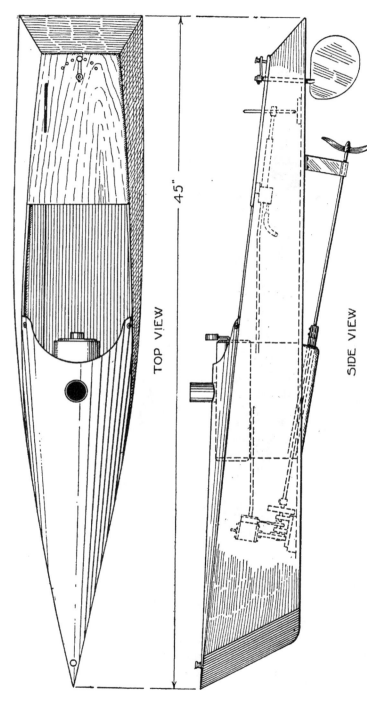

TOP VIEW

45"

SIDE VIEW

Fig. 178—Complete plan of the sharpie-type boat "Experiment"

bow, and another about, or, better still, a trifle behind center.   The bow piece and stern piece should next be made of good clean pieces of oak, free from knots.   After

Fig. 179—The boat with the deck removed.   The hand water pump is shown at the right

this preliminary work has been done, the bow piece molds and stern piece can be mounted and nailed on a board as shown in Fig. 180.

The following is needed: About 12 feet of ³⁄₁₆-inch mahogany 9 inches wide, some odd pieces of ¼-inch oak for knees and a piece of oak about 1½ inches thick for the bow. The sides are marked out and shaped from the drawing; both exactly alike. The bow piece is planed into a triangular section and rebated to take the fore end of the side planks. The stern piece is made

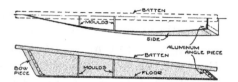

Fig. 180—How the hull of the boat is assembled

of ¼ inch oak and beveled to fit into the sides. In the actual building, the sides are secured to the bow piece by brass screws (No. 1, ½-inch wood screws) and then sprung out and the molds fitted; the transom being

Fig. 181—The Scott semi-flash boiler used on the ''Experiment''

finally fixed in place. This is fastened by angle pieces of sheet aluminum bent to fit in the angle between the sides and the stern pieces. The angle pieces are first screwed to the side planks ¼ inch forward of the aft edge of the sides, the stern piece being finally fitted into place and secured by screws through the side planks at the edge of the stern piece. The holes for all these screws should be drilled in order to avoid splitting the side planks with wood screws.

Plane off the floor edge of the side planks perfectly level. The floor is next fitted. Lay the floor board into place and mark off all around the side planks. Then take it off and rough out the shape and nail it into place with brass brads. Drill all the holes for the brads to prevent splitting. Before doing this, paint the edge of the sides with thick paint. After the floor board is fastened firmly

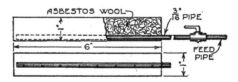

Fig. 182—The alcohol burner used to fire the boiler

into place, plane off the edge, flush with the sides. The molds can now be removed and the knees, cross-braces and engine mounts fitted. The knees are fastened by first being glued into place and then two screws are put into each leg. The after-deck is a piece of $\frac{5}{16}$-inch mahogany permanently fastened down with brackets. The remaining jobs are making the turtle deck out of a thin sheet of aluminum and the fitting of the shaft skeg, which is a piece of 1-inch square oak drilled to take the propeller shaft stuffing box. This stuffing box is nothing more than a piece of $\frac{3}{8}$-inch diameter tubing plugged at each end for about $\frac{3}{8}$ of an inch with brass and drilled the size of the propeller shaft that is going to be used. When the shaft is finally put in place, the tube should be filled with vaseline thinned out with a little oil. If the builder has been careful of the outside surfaces of the hull, he can make a natural wood finish highly polished and with a polished brass cut-water put onto the bow, a very pleasing model can be produced.

The "Speedy" engine, described in Chapter IX, and the water tube boiler, which is described in Chapter XVII, Fig. 153, would make a splendid power plant for this boat. The boiler can be fired by an alcohol tray burner, with an automatic feed arrangement which prevents the burner from overflowing.

In constructing the burner, the tray can be made about 6 inches long, 1 inch wide and 1 inch deep. At

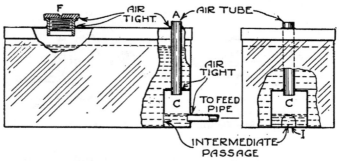

Fig. 183—The fuel tank and automatic feed for the burner

the bottom is a ³⁄₁₆-inch brass tube with pin holes. This is covered with asbestos wool, as per Fig. 182. The alcohol tank, which embodies the automatic feed device, is shown in Fig. 183 and can be built out of a cocoa can with a small chamber, C, soldered onto the side at the bottom with air tube A and intermediate passage I, as shown in the drawing. There should also be a filling plug, F, which should screw down airtight on a leather washer. The successful working of this burner depends upon the alcohol tank being perfectly airtight.

The plunger pump drawing shows this necessary attachment very clearly and needs no further explanation. The boiler feed pipe goes to the check valve on the boiler.

The propeller is built up, the blades being soldered onto a hub which is drilled and tapped for set screws. The pitch of the blades can only be determined by experiment when the boat is tried out.

The rudder is· of thin sheet brass soldered to the rudder post of ¼-inch diameter brass rod.

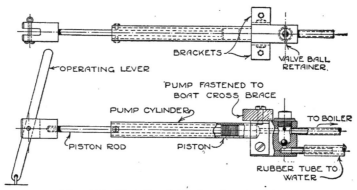

Fig. 184—Details of the hand water pump

As an afterthought, a hatch can be built in the turtle deck over the engine large enough to allow for oiling and adjusting the engine.

The alcohol burner will function better if the asbestos wool is underlaid with absorbent cotton, as this material possesses a greater capillary attraction than the asbestos and the feed will therefore be more satisfactory.

CHAPTER XXII

A MODEL SUBMARINE CHASER

Method of constructing the hull—Electric power plant and transmission—
Deck fittings.

This little craft is 34 inches long, 5½ inches beam, and it draws approximately an inch of water when loaded with driving mechanism and battery. The drive is by means of a single screw connected with a battery motor

Fig. 185—The submarine chaser complete

placed well forward. The propeller shaft housing is securely held in its proper relation to the hull by means of brackets formed by bending annealed brass strip to the proper shape and afterward filing it thin and smooth. One bracket, the inside one, is soldered to the shaft tube in order that a firm bearing may be secured at the point

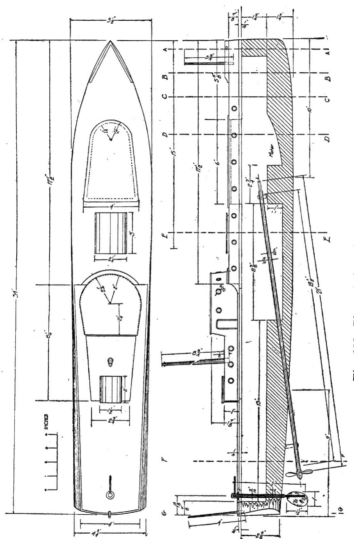

Fig. 186—Plan of the submarine chaser

where the power is transmitted from motor to propeller shaft. The coupling is the simple type comprising a short length of coiled brass spring. The outside bracket must, obviously, be left unsoldered in order that the shaft tube may be inserted through the hull.

The motor was removed from its cast iron base and supported between angles of brass bar, as shown very clearly in Fig. 192. This method of mounting makes possible the alignment of the motor shaft with reason-

Fig. 187—The stern of the submarine chaser

able accuracy. The flexible coupling takes care of whatever inaccuracy exists between the motor and propeller shafts.

The rudder is of the balanced type and of very simple construction. The profile is traced upon a sheet of brass and the blade cut roughly with snips and finished with grinder or file. The rudder post is a length of brass rod split with the hack saw sufficiently to take the rudder blade. A hole is drilled and an escutcheon pin inserted near the end. This pin is then headed over to draw up the halves of the split rod in order that the job may be neatly soldered. After this important operation has been performed, the head of the rivet may be ground or filed off and the job smoothed up with emery cloth.

The rudder post passes through a tube of brass fitted with a suitable flange at the top to provide a nice finish. The tiller is a short length of smaller brass rod inserted

Fig. 188—A view of the cabin on the chaser

as shown in the illustrations. No particular provision has been made in our model to hold the tiller in any desired position. This might readily have been done by forming a segment of a circle of brass sheeting with depressions at the desired points to engage the tiller.

Two pieces of 1¾-inch thick sugar pine form the body of the hull. They are worked out roughly to finished shape on the jig saw, formed up with the draw-knife and finished with spoke-shave and sandpaper.

The inside of the hull is hollowed out before the two pieces of pine are permanently cemented together. The lower piece is bored and gouged out. One of the photographs shows the method employed to hold the stock to the bench while the gouge is being used. The bit of stock supporting the motor and the propeller shaft bearing is left in position, the gouge being worked around it.

The two pieces are then cemented together with a mixture of white lead, whiting, bath tub enamel and japan drier. The paste is smeared liberally on *both* faces to be joined; they are then subjected to pressure by means of furniture clamps or any other convenient method. The cement will harden in three days. It is well to use a few very slender brads in joining the hull planks to assist

Fig. 189—The bow of the boat, showing the deck gun

in making the structure rigid. If the walls are brought down as thin as they should be, this task of placing brads will be a delicate one. The expedient employed was to drill down through the gunwale with a No. 55 drill to insure a straight path for each brad.

The deck and forward superstructure, if such it may be called, are integral. A piece of ¼-inch stock runs over the entire hull while the rise forward is formed by the addition of a piece of ⅞-inch stock to this thin decking forward. The deck and its companion piece are secured after being finished closely to shape in the vise. The superstructure should be hollowed out a bit if possible to decrease the weight at this critical height.

The gunwale is a length of 3/16-inch square stock running the circuit of the deck. The gunwale is painted

Fig. 190—How the boards that compose the boat are gouged out. The complete hull is shown above

gray while the deck is stained and varnished after the "plank lines" have been laid in it. These lines are merely score marks made with a scriber to imitate the lines of the usual deck planks.

The "conning tower" is of wood, hollowed out to the

thinnest possible degree and decked over with cigar box wood. The structure is made removable so that access may be had to the battery amidships.

The port holes are finished with short pieces of brass pipe cut off in the lathe with a parting tool so ground

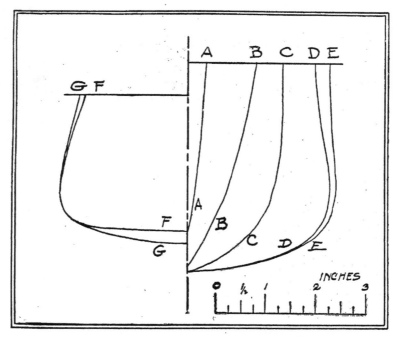

Fig. 191—Half sections of the submarine chaser hull

as to produce the desired finish on the face of each port. A bit of celluloid placed in each hole in back of the brass collar gives the desired semblance of glass. The holes in the hull and conning tower are ⅜ inch in diameter to make a snug fit for the brass collars which are forced into position after the hull has been painted.

The wireless aërial rigging and the spars are turned up from clean dowel rod. The little insulators are bits of 3⁄16-inch dowel rotated at very high speed in a drill chuck in the lathe and turned to the desired shape with a very sharp tool. The chief difficulty will be met and overcome if these two precautions are exercised.

The railings and stanchions are respectively of hard, smooth linen thread and slender steel brads. The brads are inserted into holes drilled at the proper places. Care

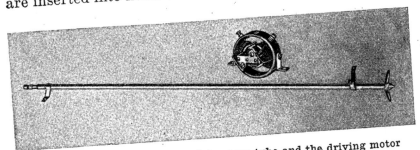

Fig. 192—The propeller shaft in the stern tube and the driving motor

should be taken in driving the brads to make them all of the same height. The thread is taken over each stanchion in a half hitch which enables the constructor to draw the lines tightly, making the appearance neat and trim. Both railings and stanchions are painted gray.

The little doors in the conning tower are of cigar box wood, carved to resemble the door and casing and secured to the tower by means of the marine cement and a few brads made from cut-off pins.

CHAPTER XXIII

A MODEL SUBMARINE WITH RADIO CONTROL[*]

Building the hull and superstructure—The radio control mechanism—
Electric power plant—Special two-point relay—Automatic apparatus.

The hull of the submarine is nearly eight feet in length and, with machinery and ballast installed, the weight is a good 175 pounds without water ballast. The hull is patterned somewhat after the Lake ship-section submersible, but the characteristic design is carried a

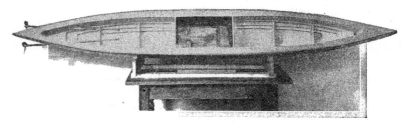

Fig. 193—The hull of the submarine completed

step farther, giving EM2 good surface riding· qualities rather than submerged speed. The bow and deck lines of the model resemble somewhat those of a torpedo boat destroyer submerged to such an extent that her deck is almost awash. This design gives ample room for machinery and controls, and it affords space in the lower hull for ballast tanks by means of which the craft may be partially submerged while at rest.

Fig. 193 is a photograph of the finished hull of white pine, with all compartments in place, and with the

[*] This book is published at a time when wireless work of an experimental nature is prohibited by the United States Government. The model maker is cautioned to bear this in mind until the order is withdrawn.

twin propellers installed. Fig. 195 is a scale drawing of the hull in plan and section.

For convenience, the boards of which the hull is made, are numbered from 1 to 12, starting with the top board or gunwale piece and finishing with the keel piece. The latter piece of board is not permanently secured in building the hull as it is to be discarded after the hull lines are developed. This board is displaced by the actual keel of lead shown in Fig. 196.

From Fig. 195, the builder can lay out full-sized patterns of the twelve planks on heavy paper from which

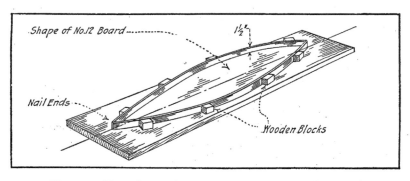

Fig. 194—How the mould is made for casting the lead keel

the designs may be transferred to the boards preparatory to sawing out. Let it be understood that the dimensions given are the finished sizes. The hull planks are to be cut somewhat oversize, say ¼ inch, to permit of working down with draw-knife and spoke-shave.

The method of procedure is first to saw out the entire twelve planks roughly to shape, nailing them with just a few brads to each other to form a more or less substantial mass of wood which bears some resemblance to the finished hull. Then, before the inside is hollowed out, the draw-knife work is started. The tools must be keen, and the most effective method is a diagonal stroke which

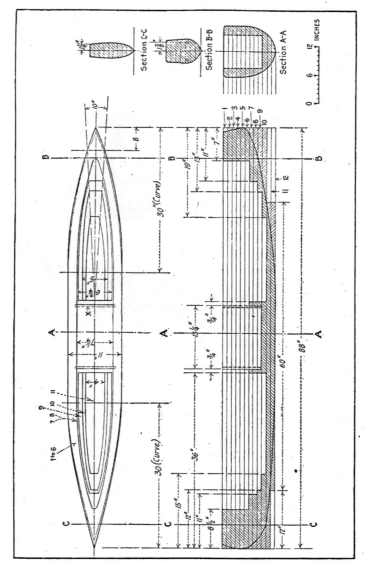

Fig. 195—Plan of the submarine hull complete, with dimensions

pares off the wood just as a jack-knife would cut it. This stroke lessens the labor tremendously and produces great execution.

The hull may be worked down to its finished lines with the draw-knife before the work on the inside is

Fig. 196—The keel of the submarine laid in place

attempted. The only part to leave for the last is the actual "smoothing down" process.

Before taking the planks apart ready for sawing out the interior of the hull, be sure to mark each plank with its number in such manner that the number cannot be readily effaced. Also mark for "bow" and "stern" to obviate possible confusion in reassembling. When this is done, take all planks apart and draw the brads. Then mark the No. 1 plank with the opening at the top which has for its most important dimension 7¼ inches in width. This is made for the benefit of the storage battery, which is of the standard automobile lighting type and of 6 volts, 80 ampere-hours capacity. A storage battery of the same type but of 4 volts and 40 ampere-hours capacity supplies the current for the controls while the larger one drives the vessel. As the width of each battery is 7⅛ inches, the dimension specified is ample to accommodate the set. The other dimensions of the opening are not so important and the opening may well be scribed directly upon the No. 1 plank, after which it is to be cut out by means of a compass saw or preferably a power jig saw. This opening serves as the guide or template for the openings in planks 2, 3, 4, 5·and 6. These openings are exactly alike.

Reference to the sectional view shows that Nos. 7 and 8 begin to form the compartments. Nos. 9 and 10 are still smaller, to care for the curvature of the lower

Fig. 197—Cross-section of the complete hull, showing the location of the water ballast compartments

part of the hull. These openings should be scribed individually on the several planks to make sure that the opening, when cut, will not pierce the wall or make it too thin in places. Fig. 193 shows very clearly how the openings become smaller as the bottom of the hull is reached.

The assembly of the hull planks, after the openings have been cut, is started with No. 1, which is placed face down on the bench. No. 2 is to be nailed to this, care being taken to "register" the planks accurately. The other planks follow in natural order. Before nailing the planks, however, attention must be called to the sealing compound which cements the boards together and renders the hull impervious to water. The cement is made as follows: To one pint of the best bath tub enamel add one pound of good white lead, stirring thoroughly until the lead is taken up by the enamel. Then add ½ pound of ordinary whiting (powder), mixing until the mass has assumed the consistency of thin, smooth putty. Add to this ¼ pint of good dryer to make the cement set quickly.

Apply the compound to *both* planks where they are to come in contact. A brush will serve notwithstanding

Fig. 198—The storage batteries used on the submarine

the thickness of the mixture. After making sure of the register, nail bow and stern with a single brad to hold in place. Then start at one end and nail with 1½-inch brads for this, the starting plank. Place brads not more

Fig. 199—The storage batteries in their compartments

than 2 inches apart throughout the entire length of the hull.

When the second plank is secured to the first, wipe the cement from the seam where it has been squeezed out and paint the surface of No. 2 and of No. 3 ready for register and nailing as before. Repeat with No. 4, but when that portion of the hull is reached where the propeller shafts are to pass through, refrain from placing nails in No. 4 and in No. 5 at the critical places where the bit would have to pass in boring the propeller shaft holes. This location is clearly indicated in the drawing and the photographs.

When the hull has been assembled up to the point where the No. 12 or keel plank is to be laid, nail this one on with a dozen brads but do not put any cement in the

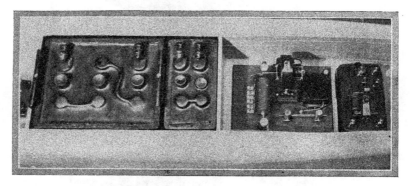

Fig. 200—The storage batteries and central control mechanism

union. As stated before, this plank will ultimately be discarded.

The hull may now be wiped clean with a cloth slightly moistened with turpentine and the cement permitted to dry for at least two days. After this, the spoke-shave may be used freely to true up the lines, removing the inevitable "bumps" left by the draw-knife. The final finish is with sandpaper. The priming coat of paint may then be applied. Good ship's paint in battleship gray is used. To this is added a quantity of pure white lead to lighten the shade somewhat. The first coat is applied after a cradle has been made for the hull. Here and there could be seen a depression or a "bump."

Fig. 201—The driving motor with transmission gears mounted on the frame

After the first coat is dry, the spoke-shave is used freel
taking down the high spots. The nicks and depression
are filled with the same compound used in the assemb
of the hull. The mixture is made thicker, however,
the addition of a double quantity of whiting. Th

"putty" is different from the ordinary variety in that it "sticks" wherever it touches (the hands are no exception) and it dries as hard as stone. After a day of drying, the putty is still soft enough to permit of smoothing up with sandpaper, and when it is finally covered with

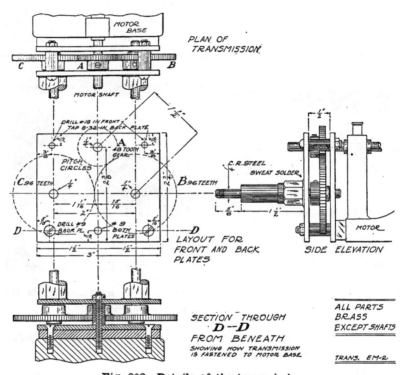

Fig. 202—Details of the transmission

paint, the imperfection is absolutely invisible. The white spots on the hull in the photographs indicate where the depressions have been filed.

When the hull paint is bone dry, the work on the interior may be started. All "whiskers" are removed with coarse sandpaper, which will smooth out the saw marks to a great extent. The compartments shown in the lower hold in Fig. 193 are merely to keep the water from surging back and forth when the craft is in motion. These bulkheads are well perforated with holes and are secured with brads. The storage battery compartment is made absolutely watertight. The drawing gives the details, the dotted lines indicating the compartment wall of ¼-inch whitewood. The cement is freely used in assembling this compartment, the walls being literally plastered with it. When the cement is bone dry, the battery compartment is given three consecutive coats of some

Fig. 203—The driving motor mounted in the hull and connected to the propeller shafts through spring couplings

good acid-resisting compound obtainable from a large electrical supply store.

The entire interior of the hull is next to be plastered with the cement, made somewhat thinner through the addition of more enamel. The mixture should be worked well into all corners and cracks until every corner is rounded out with the paint. The water compartments below should receive an especially liberal dose of the mixture.

While the interior of the hull is drying out, the builder may turn his attention to the construction of the keel which is of lead.

The nicest way to make it is to take the No. 12 plank and, using it as a pattern, make a sand mould for the lead casting. If this is deemed impracticable, make a mould as shown in Fig. 194, using the No. 12 plank to give you the profile. Take this mould directly to a manufacturer of lead pipe or plumbers' supply house and have them pour the lead. The keel should weigh 75 pounds, and the simplest way to gauge this is to place the mould on the scales and pour until the desired weight is attained. The keel is about an inch thick. Its shape and the proportion it bears to the hull is shown in Fig. 196. The keel is secured to the hull by means of a quantity of flat head brass machine screws passing through the No. 11 hull plank and into tapped holes in the keel. All cracks, holes and other blemishes are filled with the cement, and after painting the keel looks as though it had grown on the hull.

The ballast tanks in EM2 serve a twofold purpose. Primarily they are intended to partially or completely submerge the submarine while she is at the "wharf." Their secondary purpose is to "trim" the craft when she is placed in the water for the first time. The necessity for this latter procedure will be obvious to every veteran model builder. He will know from experience how nearly impossible it is to so distribute the load inside the hull so that the craft will float on an even keel when first she is launched. The tanks enable the operator to trim the submarine both fore and aft, causing her to assume not only the desired depth in the water, but also to overcome any pronounced tendency to ride lower by the bow than the stern, or *vice versa*.

As described previously, the "tanks" are built into the lower part of the hull. Theoretically they should be lined with zinc or copper. In practice, however, it has been found that this is unnecessary. A very liberal use of good waterproof paint over the interior of the hull serves the purpose without the need for a very complex piece of sheet metal work such as the tank would represent.

The tanks are completed through the addition of a decking of ½-inch wood thoroughly coated with the paint or cement described previously. This decking is to be literally cemented in place first of all and then, to add strength, it is nailed every inch or so to the step in the hull shown in the illustrations. Fig. 197 gives a section

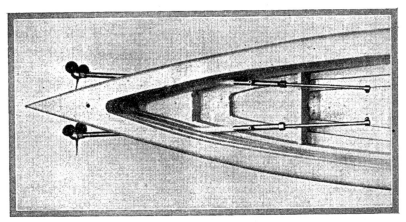

Fig. 204—The stern of the hull, showing how the propeller shafts pass through

through the hull with the tanks completed. The water is kept from surging back and forth by means of bulkheads which divide the tanks into compartments represented by W in the drawing. These bulkheads are perforated so that the water entering the openings indicated in the bottom of the hull can find its way readily into all compartments.

The intake holes through the keel are left wide open at all times. The water is permitted to enter by opening air cocks on the superstructure of the submarine; as the air in the tanks is permitted to escape through these cocks, the water enters. To discharge the ballast, compressed air is sent into the tanks through the cocks, thus driving the ballast out. This operation is the only one not performed through the medium of the radio control. It could be done, of course, while the craft is running, but the delicacy of the adjustment and the tendency of the craft to plunge to the bottom under such circumstances render such a plan inadvisable.

The details of the air cocks and their location will be taken up later after the superstructure has been covered. To give such details at this point would necessitate repetition later.

The storage battery is divided into two units; one is for the motor which drives the craft, while the other is for the controller and the various control devices. The motor battery is a 6-volt 80-ampere-hour automobile lighting battery, while the control battery is of 40-ampere-hour capacity at 4 volts. The dimensions given in the hull drawings in the preceding issue are suitable for a number of standard automobile batteries on the market. See Fig. 198.

The battery compartment has been described. Suffice it to say here that no attempt should be made to fasten the battery in its place. The two units should be easily removable for repairs or for charging in the event that the whole model cannot be taken to a place where current is obtainable. See Figs. 199 and 200 for views of the battery in the hull.

The motor selected for EM2 is of a standard type obtainable in the open market. The motor has practically all of the desirable features found in commercial machines of large size. Radial gauge brushes, mica insulated commutator, adequate oiling facilities, form wound coils, and a liberally designed frame, are among its excellent features.

The transmission has been added solely to permit the

motor to run at relatively high speed with moderate propeller speed. The arrangement of the gears is such that a speed reduction of 2 to 1 is obtained with the shafts for the propellers running in opposite directions. This, of course, necessitates the use of a left- and a right-hand screw.

The motor with its transmission is well shown in Fig. 201, which gives a front and rear view of the power plant. These photographs, together with the detail drawing,

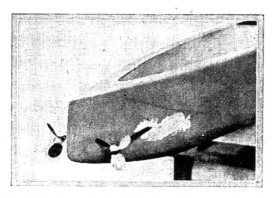

Fig. 205—The stern of the hull, showing propellers in place

Fig. 202, should make clear the entire scheme of the transmission.

The reader will note that a 48-tooth brass gear is secured to the motor shaft. This is a stock 1-inch gear of 48 pitch. Meshed with this gear is a 96-tooth gear, 2-inch diameter, mounted upon one of the transmission shafts. Meshed with this gear is a second 96-tooth gear mounted upon the second shaft. The layout drawing in Fig. 202 will make this clear. The gears are represented by pitch circles. The operation is obvious. The motor gear A turns the first large gear B, which in turn operates the second large gear C in the opposite direction.

While on the subject of transmission, it may be well

to suggest how the bearing plates are laid out and fitted to the motor frame. The first operation is to locate and drill the holes in the upper part of the plates, drilling clearance in the front plate and tap size for 8/32 in the rear plate. Place screws in the holes to grip the plates together. Next locate the center hole at the bottom and drill No. 9 to clear a 10/32 screw, which may be inserted and held with a nut. The plates may then be filed up square and true to each other. All holes will thereafter be drilled through *both* plates at the same time to insure alignment.

The ¼-inch hole for the motor shaft will be next in order. This hole is located near the letter A in the layout drawing. When the hole has been drilled, remove the 10/32 screw from the center hole at the bottom of the plates and slip the motor shaft through the hole. Clamp to the motor frame with the distance piece shown in all views between plates and frame. Square up with motor base and run a No. 9 drill through the lower hole to spot into the motor base. Follow with tap drill for 10/32 and tap out.

After this, remove the plates and replace clamping screw in the lower hole. Next lay off the center for the gear B by scribing with dividers set to 1½-inch radius from center of motor shaft hole. Strike a vertical line ¹⁵⁄₁₆ inch to right of center and spot for the ¼-inch hole for B gear shaft. Now set dividers to 2-inch radius and swing from the B gear hole to the left on a horizontal line. This will give the spot for the C gear hole. Drill B and C holes with ¼-inch drill, making sure the drill is ground true so as to insure that it will not cut oversize. The plates may now be separated and the rear plate replaced on the motor shaft with a short 10/32 screw temporarily placed in the center hole to clamp the plate to the motor base. The holes for the flat head countersunk screws may then be drilled and tapped and the plate permanently secured to the motor frame. If the work has been done carefully the motor shaft will turn without any bind when the plate is secured.

The bearings for the transmission shafts next require attention. These are turned up from ¾-inch hexagon brass stock and sweated to the front bearing plate. For this operation, a ¼-inch wooden dowel was placed in the plate and bearing holes to line them up, and the surface

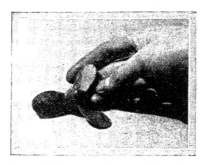

Fig. 206—One of the propellers used on the submarine

of the plate and the under side of the bearing coated with soldering paste. A very hot soldering copper was then used to flow solder around the joint, care being taken to see that the solder actually sweated into the joint between the bearing and the plate. The plate was set up on three brads to keep it away from the bench for this operation. When the job cooled, the bearing plates were found to be in perfect alignment, thanks to the dowels, although the latter were badly charred at the point of union. An oiled rag run through the hole cleaned the bearing surface of the trace of carbon left by the burned wood.

The A gear is secured to the motor shaft with a pointed set-screw, while the B and C gears are fastened to their shafts by means of brass escutcheon pins driven into holes drilled through gear boss and shaft. The ends of the transmission shafts are threaded ¼-20 to take the steel springs which provide flexible couplings to the propeller shafts.

All that remains is the assembly. The spacing collars are self-explanatory in the drawings. The reader will note that the 10/32 screw temporarily placed in the rear plate is displaced by a long stud, which passes through both plates with a spacing collar or sleeve between. These sleeves were made by cutting off the desired lengths of ⅛-inch brass pipe in the lathe. An additional feature not shown in the illustrations might well be a simple oilcup in each bearing of the transmission.

Fig. 203 shows the motor and transmission mounted in the hull. The method of coupling shown has been simplified and improved upon by removing the shaft hangers and substituting long, tightly wound steel springs of the door-spring variety. So stiff and so tightly wound are these springs that they afford an almost perfect flexible coupling to the propeller shafts. The ends of the springs are forced on the threads of the transmission and propeller shafts, where they will stay put until removed by main force. The propeller shafts were cut off close to the shaft housing shown at the extreme left in Fig. 203, and much more clearly in Fig. 204.

Little need be said of the shaft housings. They are merely lengths of ⅛-inch brass pipe fitted with a standard brass cap at either end. The caps are of course drilled for the shaft. The space between the end of the pipe and the cap is ample for the lampwick and tallow packing necessary to exclude water. The size and shape of the propellers is well shown in Fig. 206, where a hand is included to give an idea of the proportions. The average builder will want to take advantage of a stock propeller rather than make patterns and have just a single casting made from each. The relation of the propellers to the hull is shown in Fig. 205. They are secured to the shafts by a simple screw thread and pinned to prevent loosening.   This threading operation was easily accomplished by chucking the propeller in the three-jaw chuck, facing off, centering, drilling and tapping ¼-20.

The steering rudder is well shown in Fig. 207, as it appears beneath the hull at the stern of the model.   Fig. 208 gives a glimpse into the hull, showing the rudder

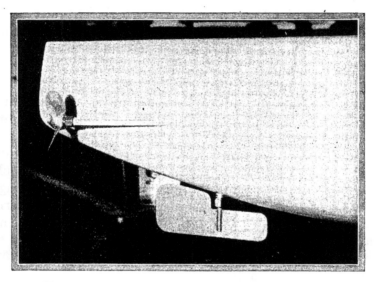

Fig. 207—The stern of the submarine hull, showing one of the propellers and the rudder

mechanism in position.   Fig. 209 shows the rudder mechanism in parts, giving an idea of the proportions of each piece to the others.

For the constructural data, the reader is referred to Figs. 210, 211, and 213, which are working drawings of the several parts of the control.   The remaining illustrations, Figs. 212 and 214, show the complete mechanism installed within the hull.

For an explanation of the working principle let us

to Figs. 210, 211, and 213, with an occasional glance at the photographic views. The rudder proper is of sheet brass sweated into a saw-cut in the end of the rudder post.

Fig. 208—The rudder control mechanism

The latter passes through a brass pipe sleeve or housing fitted with packing glands at either end.

To the top of the rudder post is fitted a brass disc having a shallow groove turned in its periphery to carry

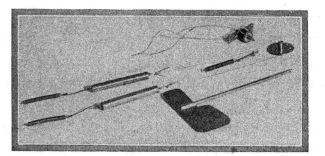

Fig. 209—The parts of the rudder control mechanism

the "tiller cord" or its equivalent in this case. The construction and assembly of post and disc is clearly shown in the detail drawing, Fig. 210. The hub of the disc is sweated with solder to the disc. The projecting arm of the post is threaded into the latter and tightened by means of the lock nut shown. The function of this pro-jecting arm is to carry the central spring, which helps keep the rudder in a neutral position.

In the disc there are two little countersunk holes. See Detail B in Fig 211. Each of these holes is 25 degrees

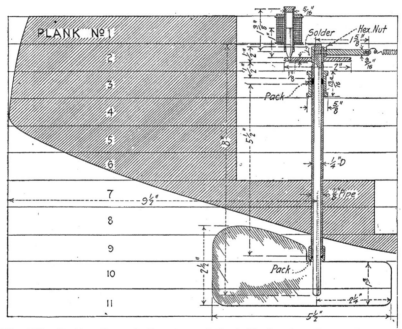

Fig. 210—Section through the stern of the hull, showing the arrangement of the rudder mechanism

to one side of the neutral position of the rudder. Mounted directly above the disc, see Fig. 210, is a small solenoid in which a plunger of iron slides freely. The lower end of this plunger is conical in shape to fit the countersunk holes in the disc.

The action of the mechanism is as follows: When the tiller cord is pulled to the right (by solenoids to be de-scribed) against the tension of the neutralizing spring, the disc moves beneath the solenoid plunger until the latter drops into one of the holes. Here the rudder will stay until the next operation, which sends a current through the little solenoid, releases the rudder, sending it back to the neutral position. Likewise the left-hand pull of the rudder serves to engage the solenoid pin with the second hole.

The big feature of this mechanism is the fact that current is used for an instant only—just long enough to pull the rudder to port or starboard. It remains in that posi-

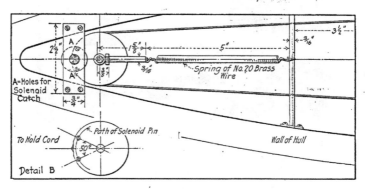

Fig. 211—Plan of the stern and rudder mechanism

tion until released, although no current is used in the retaining operation. The net result is that we can use a comparatively large current for the actual pull of the rudder, making the operation positive and trustworthy, and still maintain the essential economy of current consumption.

The tiller cord has been referred to. This is a length of the round type of shoestring, strong and very flexible, attached to the disc by the retaining screw shown in Detail B, Fig. 211. This cord, running in the groove in the disc, passes on to a large solenoid on either side of the hull. The photographs and the drawings, Figs. 211, 212, 213 and 214, will make this clear. The plunger of each solenoid passes entirely through the coil in order that tension springs may be attached to the rear end of each plunger to assist in keeping the rudder in a neutral position and to keep the tiller cord tight.

The small solenoid used to retain the rudder in the desired position is wound with No. 24 wire, enameled, in layers 1 inch long upon the brass tube of the mechanism. The careful workman can get on 45 turns per layer and

Fig. 212—The complete rudder mechanism, showing the solenoids

the winding should be about 18 layers deep. This winding is designed for use with the 6-volt storage battery.

The large steering solenoids are wound with No. 16 D. C. C. wire, 68 turns per layer, and 8 layers deep to each coil. The brass tube upon which the solenoid is wound should be a sliding fit over a piece of ½-inch Norway iron rod made perfectly smooth. There must not be any friction in this sliding fit. If Norway iron cannot be obtained, cold rolled steel may be used, but the pull is lessened appreciably.

The mounting for the solenoids is of hardwood. In mounting the coils within the hull, care should be exercised to see that the plunger of the solenoid has a straight pull on the cord rather than one at a slight angle. It will probably be necessary to block out the rear end of each solenoid base with shims of wood to effect this.

The current density in the solenoids is comparatively high. The small coil pulls about 5 amperes, or enough to heat it disastrously if the current were permitted to re-

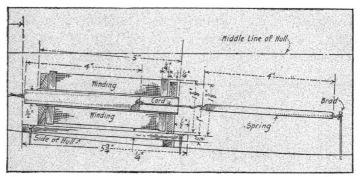

Fig. 213—Details of one of the controlling solenoids of the rudder mechanism

main on. The large coils draw 15 amperes apiece at 6 volts. This amount of current produces a pull that will draw the plunger out of one's fingers nine times out of

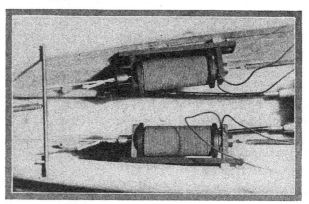

Fig. 214—Close-up view of the rudder solenoids in place

ten, providing the plunger is inserted three-quarters of its length when the current is applied. This high current density is not objectionable, however, in view of the fact that it is applied for but one or two seconds at a time.

In the radiodynamic control of a distant model of some sort, there are a number of practical considerations involved aside from the actual problem of how to do the trick. For instance, the distance over which control is to be effected is most important. The difficulties increase amazingly as the distance between transmitter and receiver increases. Another point to consider is whether

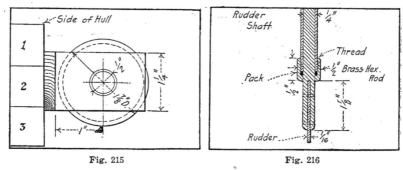

Fig. 215   Fig. 216

Fig. 215—Dimensions of the solenoid supports
Fig. 216—Section through the rudder post

the control must be selective or whether a progressive series of evolutions, each dependent upon the pressure of the radio key, is all that the builder desires.

One thing that is absolutely essential in any system of this kind is some means of determining just what is going on inside the hull of the model when the key is pressed. True, the craft will respond, but we must know beforehand just what is going to happen at the instant of pressure of the key.

There are two ways of doing this. One is to have a series of colored lights on the deck of the model, each color representing some distinct evolution of the craft. A chart on shore tells the operator what each color means so that he can guide his signals accordingly. The other method involves the use of a clock-work mechanism to operate the control fingers, with a synchronous mechanism, turning at exactly the same rate of speed on shore in front of the control operator. By observing his clock the operator can determine exactly what contact finger is

Fig. 217—The central distributor for the three-circuit controller

in connection at any given time and can send his signals accordingly.

The colored-light signal device is admirably adapted for use with models where the control need not extend more than a few hundred yards at the most.

The system consists essentially of a mechanism the function of which is to periodically open and close a certain number of circuits. The colored lights, each color representing a circuit, tell at all times the position of the contactor. The function of the radio impulse is to complete the operative circuit to any given device in the submarine at just the desired moment. In other words, the operative circuits are all open, as the contact finger moves from one point to another, until such time as a radio signal comes in. This serves to close a relay, sending current through the desired solenoid or motor or whatever happens to be the operative mechanism of the circuit in question.

In the design for the distributor or controller-proper, there are a number of considerations. If space is at a premium, and a very large number of controls is deemed necessary, the type shown in Fig. 218 is better. This device is essentially an instrument switch having a double row of contacts and a contact arm which is continuously rotated by means of an electric motor mechanically connected by means of worm gearing. The small rheostat is to control the motor speed.

In this controller or distributor, the operative circuits are taken from the outside row of contacts while the lamp or signal circuits are taken from the inner row. The one advantage of the two rows is found in the fact that but a single white lamp need be used to indicate the "off" position of the contact finger following each colored flash.

The coherer and decoherer are mounted upon an upright wooden piece at the end of the controller base in such a position that the coherer tube is readily accessible for adjustment through a hatch in the deck of the submarine.

The distributor shown has twelve contacts, which means that it will control six complete circuits or evolutions; that is, six on and six off. Thus the model can be made to start, stop, turn to right, turn to neutral, turn to left, back to neutral, fire a deck gun, flash on a searchlight, turn it off, blow a whistle, discharge a torpedo, fire

Fig. 218—The central distributor for the six-circuit controller

a signal rocket. Or, in lieu of two of these evolutions, it can be made to submerge and rise with the diving planes fore and aft. The point to remember is that any operation that is performed in an instant, such as firing a gun, can be done without the expenditure of the second contact to turn the current off. All of the devices such as rudder solenoids, driving motor, etc., in EM2 are so designed as to keep on operating until a second impulse turns the current off.

The distributor shown in Fig. 217 is somewhat easier to build, but it does not lend itself so readily to a large number of circuits. In this device the cylinder carrying a series of contact studs, each contacting in turn with a brush or finger, is revolved slowly with the usual worm gear and motor drive. This distributor has six contacts, which means that it is good for three evolutions: start motor, stop motor, turn right, turn back to neutral, turn to left, back to neutral. This is all that would he required of the average torpedo and for simplicity of construction and control, this form is splendid.

Still a third form may be used and this has some pronounced advantages over the others. It consists of a segmented ring of metal, each segment representing a circuit, around which a brush travels slowly. The advantage here is that it gives the operator the maximum of time in which to read and respond to his signals.

The large wiring diagram will make all connections clear without many words of explanation. In reading the diagram, the builder is advised to trace with a pencil each circuit, keeping in mind the position of the contact arm at the time. For instance, to start the motor, he will see that the control current is sent through a two-way switch that will stay in whichever position it is drawn by the magnet on either side. To stop it, the current is sent through the opposite magnet for an instant to draw the lever over.

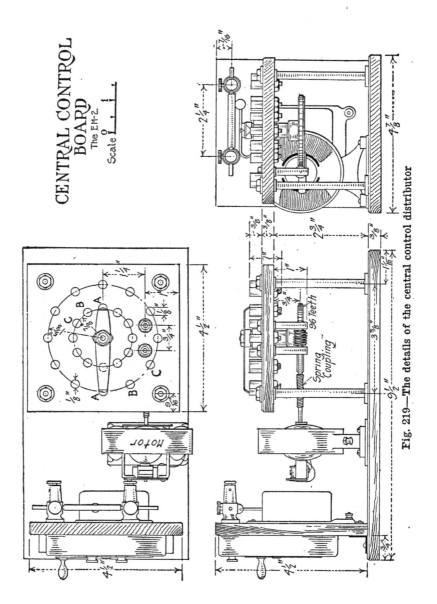

CENTRAL CONTROL
BOARD
The EM-2.
Scale

Motor

96 Teeth

Spring
Coupling

Fig. 219—The details of the central control distributor

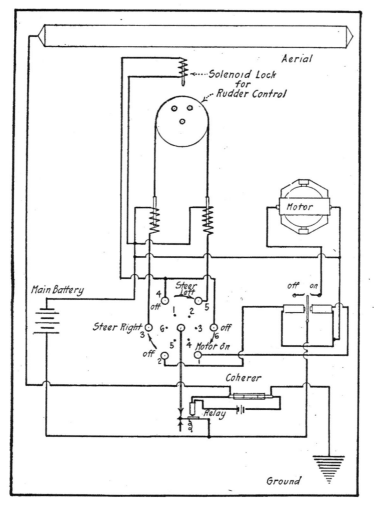

Fig. 220—Wiring diagram for the three-control distributor. Each circuit and control is connected with a colored light

The coherer is the inherent weakness of any radio control system. For simple model work the old fashioned coherer with silver plugs and a half-and-half mixture of nickel and silver filings is about as good as any. The plugs should be at least half an inch apart in the glass tube and the intervening space loosely filled with the filings mixture.

One problem in the construction of the Model Submarine EM2 was to provide a means whereby the current operating the main driving motor could be turned on or off by the control device. The switch controlling this circuit, of course, had to be of such a construction that the contacts would remain in the desired position until the next impulse came. Furthermore, from the economic standpoint it was essential that no current be used to maintain the switch contacts in either position, the impulse serving merely to bring the contacts to the on or off position as the case might be.

Some experimentation resulted in the relay switch shown in the photograph, 222, and the drawing, 221.

Essentially, the switch consists of an arm of soft steel, pivotally arranged between the poles of two magnets, and above this armature two springs of phosphor bronze so disposed as to contact with the end of the armature as the latter is drawn to one side or the other by means of the magnet through which current is sent.

The magnets are of the standard instrument type of 2.5 ohms resistance each. These magnets are sold at such a low price in the open market that it is cheaper and more satisfactory to purchase them outright than to construct them. Each magnet is mounted upon the upright of a piece of brass angle as shown in the illustrations. The distance between armature and pole piece of the magnet should not be greater than $\frac{3}{32}$ inch to insure reliable operation without the expenditure of an inordinate amount of energy in the act.

The armature is a piece of cold rolled steel (unless wrought iron is available) of $\frac{1}{4}$-inch by $\frac{1}{2}$-inch section. It is drilled for the pivot at its lower end and this fit should be a good one. The armature should swing freely back and forth but a loose and slovenly fit cannot be tolerated as it will lead to trouble.

The spring strip contacts are mounted upon brass angle pieces and their adjustment is shown by the photograph infinitely better than a thousand words of explanation could describe it. This adjustment is possibly the only "ticklish" part of the job. Too much pressure will prevent the armature from being drawn to the reverse

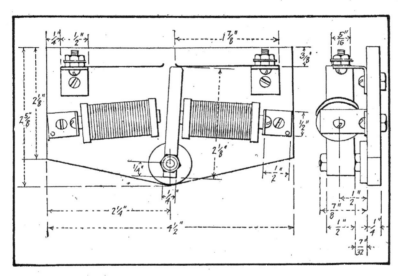

Fig. 221—The two-point relay switch

position. Too little pressure will not hold the armature in the desired position against the inevitable vibration of a model boat hull. Of course the design of the switch is such that gravity aids in keeping the armature in either one position or the other as, when it shifts from one side to the other, the center of gravity shifts with it. When the correct adjustment is obtained, the armature will click over from one position to the other the instant the current is applied and it will stay in that position until current is sent through the other magnet.

The superstructure of the model is so designed as to permit of ready access to the machinery within through the openings which are covered with simple hatches of wood secured by means of brass nuts on screws passing up from beneath.

A complete drawing of the superstructure and decking is given in Fig. 224, which shows a plan and a side

Fig. 222—The relay switch assembled

elevation in section. Note that the superstructure is built up of two thicknesses of $\frac{7}{8}$-inch material and one of $\frac{3}{8}$-inch stock. The latter is secured after the former are worked down to finished shape. In this thin piece of lumber are placed the brass machine screws which hold the nuts for the hatches. All dimensions are given in the drawings.

In placing the superstructure on the hull, and, likewise, in nailing the two pieces of stock together, the white lead and whiting cement is to be used plentifully. The cement is to be applied to the stock on both surfaces before nailing together.

To counteract possible warping, the superstructure should be secured with both screws and nails. A flat-head screw placed every 8 inches or so will draw the superstructure right down to the hull, squeezing out the cement which can be scraped off when nearly hard, leav-

Fig. 223—The submarine with the deck removed

ing a perfect union between hull and superstructure. As the decking of thin wood is placed on the superstructure after the latter is securely fastened, the screws and nail heads are covered.

The wiring and the installation of the machinery were done after the superstructure had been finished. The object of this was to insure that the interior fittings might be removed after the deck was finished.

The greater part of the wiring was done with No. 14 standard rubber covered and the balance with No. 18

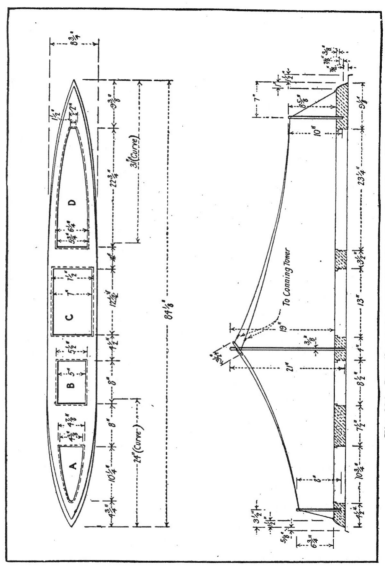

Fig. 224—Details of the superstructure for the submarine

fixture wire. The insulation on these conductors is so good that little care need be exercised to keep them apart. They may be "fished" beneath the deck and held under

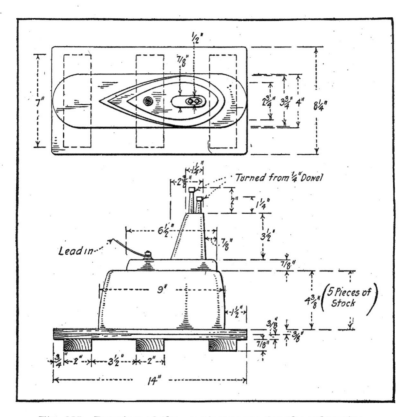

Fig. 225—Drawing of the conning tower for the submarine

staples with impunity. On all circuits where the current is heavy, the larger wire was used.

As a measure of safety, 10-ampere fuses should be inserted in the leads from the large storage battery and 5-ampere fuses in those from the small control battery. This protection is really very important, as the current from either battery on short circuit might easily reach a hundred amperes or even more. Incidentally, such a discharge would ruin the battery. The heaviest current drawn is 15 amperes; however, this is only for an instant and the fuses specified will hold it easily for the second or two it is on.

As an additional precaution, it is well to provide switches just inside the battery compartment in order

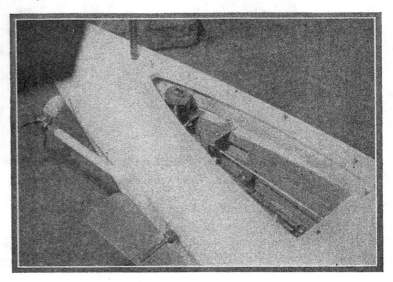

Fig. 226—The stern of the craft, showing the rudder release solenoid

that the two batteries may be totally disconnected from the mains when the model is not in use. In the photograph showing the battery compartment, Fig. 231, the switch at the left is to disconnect the main battery, while the one at the right starts and stops the control device motor. When these switches are open, no current can flow through the circuits, even though the controller might be left in an "on" position.

EM2 has controls as follows: turn to right, neutral; turn to left, neutral; start motor, stop motor. The diving plans are manually operated; that is, they are tilted to the diving position by hand. Should any reader, in building this model, choose to add the radio control, he

Fig. 227—The driving motor, with relay switch mounted in place

has but to construct a mechanism similar in principle to that employed in the control of the rudder.

In order to use the central control device described, the contacts have been grouped into pairs by bridging

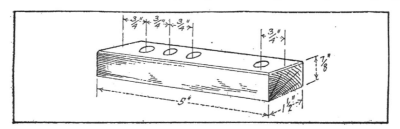

Fig. 228—The block upon which the signal lights are mounted

across adjacent studs so that each contact is repeated as the arm travels around. The leads from the contacts are of flexible lamp cord brought up to screws and washers along the side of the opening in the superstructure.

The relay switch, which serves to turn on and off the current to the driving motor, was described previously.

The drawings give all details of the conning tower and deck fittings. The aërial is carried by a tall central mast and it inclines toward a shorter mast at either end of the craft. A thoroughly dry dowel boiled in paraffin provides a good mast of high insulating qualities. So effectual is this treatment that no insulators are found

Fig. 229—The controlling mechanism exposed by the removal of the forward hatch

necessary. It is admitted, however, that the true model maker will not be content with such an arrangement. For him the employment of tiny insulators turned from hard rubber or black fiber rod of about ¼ inch diameter is necessary.

The lead-in is taken from the center of the aërial and brought down to the contact on the deck of the conning tower. From this point, a piece of rubber covered lamp cord leads to one side of the coherer. The other side of the coherer is connected with a heavy wire that leads down through the hull and makes electrical connection with the lead keel.

The ballast tanks in the lower part of the hull are of sufficient size to hold water to completely submerge the model. A piece of brass pipe leads from the machinery deck both fore and aft to the outside of the hull where each pipe terminates in a small pet cock. When these cocks are opened, the air in the tanks is permitted to

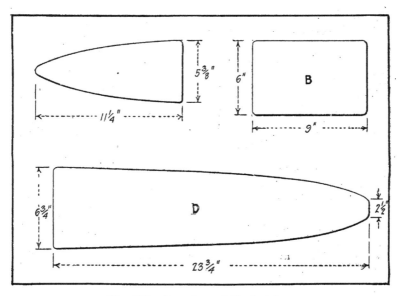

Fig. 230—Drawing of the hatches

escape, with the result that water enters. When the model is "trimmed" by the bow and stern, the pet cocks may be closed, thereby preventing the entrance of more water. To discharge the water ballast, air pressure is applied to the pet cocks by means of rubber tubing from a tank.

The coherer in the control device is sufficiently sensitive to respond to the impulses sent out by a spark coil of the standard "2-inch" wireless type, using the obsolete method of connecting the antenna and ground across the spark gap.

The maximum distance for fairly reliable operation was found to be about 100 yards. A tuned transmitter, tuned to the exceedingly short natural wave length of the diminutive submarine aërial, would increase the distance as would also a transmitter of greater power.

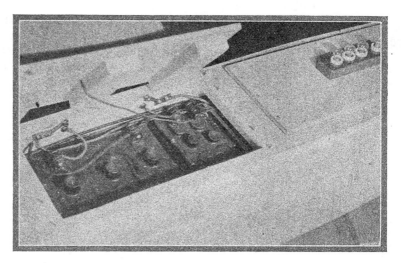

Fig. 231—The central hatch lifted, showing the storage batteries. The signal lamps are shown at the right

The adjustment of the coherer is a tedious task, but, when once attained, it is reasonably permanent. A mixture of about equal parts of pure silver and nickel filings is used, which must be quite free from grease. The best working distance for the coherer plugs was found to be between ½ and ⅝ inch with the space between practically filled with filings.

CHAPTER XXIV

A MODEL CRANE

The design of the crane—Its construction—The electrical driven hoist.

When the amateur constructor interests himself in model crane construction, he delves into a field that has a wide range of interesting apparatus. There are many types of cranes, any of which, if constructed, will well repay the builder, as the amount of instruction and

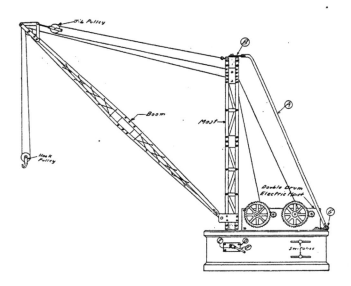

Fig. 232—The plan of the electric derrick crane

amusement derived from them is great. Of all the many kinds, the most common is the derrick or jib crane. This type is seen in lumber yards, stone quarries, steel mills, and especially in construction work.

It consists of a mast, Fig. 232, to which a boom or jib

is attached so that it may swing up or down. A movable pulley with a hook is run from the end of the boom. These two elements, the boom and the hook, are separately controlled by cables, which are wound on winding drums.

In this model crane the double drum electric hoist, which is described in the following chapter, is used. Miniature structural iron is used in the mast and boom

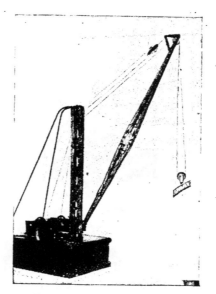

Fig. 233—The derrick assembled

of the derrick, and its use did not bring forth as many obstacles as contemplated. This feature can be used in many other models, as it is easily and cheaply constructed.

It merely consists of angle tin held together by means of very small rivets. Iron round head rivets, ⅛ inch in diameter by ³⁄₁₆ inch long, are used. These are the smallest procurable and have to be sawed to about ⅛ inch long. They are driven cold on account of the difficulties

encountered in fitting and holding same if they are driven
hot. The round head is left exposed.

The boom and mast are made in this manner: While
sticks of wood will suffice for the pieces, the realistic appearance of this material is seen by the photographs,
Figs. 233, 234.

The angle iron or tin used as the sides of the mast
and boom is obtained from a tinner who has access to a

Fig. 234—Rear view of the derrick

bending machine. Four each of twenty- and thirty-inch
strips are required. Holes are bored in the long angles
which are used in the boom, as indicated in Fig. 236.
These holes are bored identically on both sides of the
angle. Four web plates (B), Fig. 237, are cut and drilled
as shown and riveted to angles. The ends are then bent
down to the one-half-inch square brass end blocks (E)
and (F). The angles are held against these blocks by
tying a cord around them. Holes are then drilled through
the strips and blocks with a No. 28 drill and then tapped
with an 8-32 tap. . One-quarter-inch, 8-32, round head
screws are screwed into these holes, holding the strips
in place as shown in the assembly of the boom, Fig. 235.

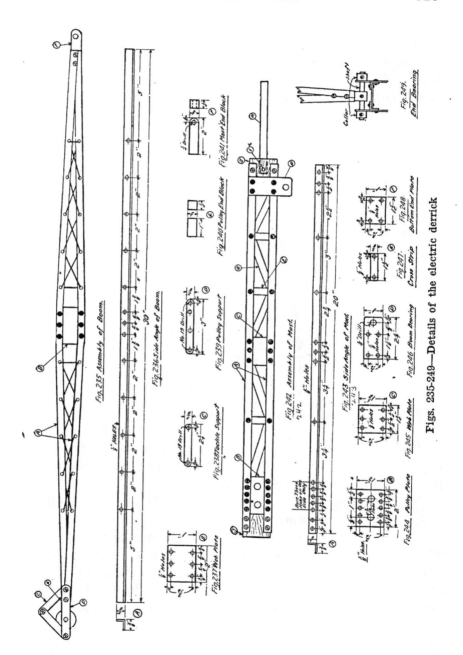

Figs. 235-249—Details of the electric derrick

A rigging, made of strips (D, E and C), has to be made so as to provide clearance for the jib pulley when the boom is in a high or low position. Details of these so-called supports are shown in Figs. 238 and 239. Four tackle supports and two pulley supports are required.

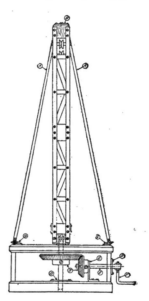

Fig. 250—The back of the derrick

The cross bracing seen in the assembly of the boom, Fig. 235, is merely No. 16 iron or copper wire, woven through the holes bored for this purpose.

The building of the mast, Fig. 242, is more difficult than the boom, as more riveting has to be done. The side angles of the mast are bored, as shown in Fig. 243. The web plate (B), Fig. 245, should be put on first. It will be noticed that no detail has been given for the diagonal strips (G), as their length and shape of ends vary. While dimensions are given for the location of rivet holes in the strips and plates, the writer finds it more advantageous to bore the holes in the angles only

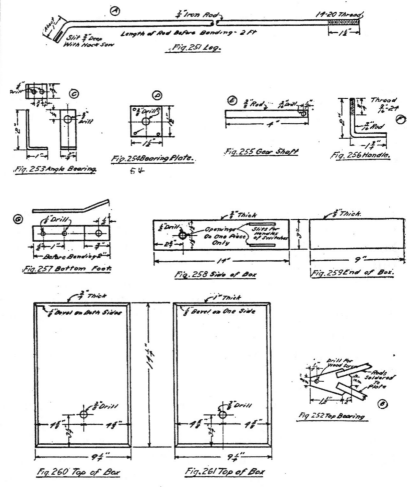

Figs. 251-261—Details of the derrick

and then cut the strips to length and locate the holes in the strips by means of the holes in the angles. It will be noticed also that where two or more strips come together only one rivet holds them together. It is best to build up the two separate sides complete and then fasten these sides together. This facilitates riveting. Only two pulley plates (B) and boom bearings (D) are required, and they are riveted on the same opposite sides.

When the sides are all fastened together, square wooden blocks (J), Fig. 242, are fastened at each end by screwing the sides to the wood blocks. The block at the bottom or the end nearest to where the boom swings has a rod (H) 5 inches long held by a pin (I). This rod (H) is a shaft for the bevel gears (J), Fig. 250. Its diameter is dependable upon the size of the hole in the bevel gears.

The boom swings at the foot of the mast by placing a shaft, ¼ inch in diameter by 2¾ inches long, through the end block of the boom and is fastened to the boom by a screw bored through the end block and shaft. Collars are provided at the end of the shaft. Fig. 249 shows this very clearly.

The mast is mounted on a box-like base in which bevel gears are mounted for turning the derrick and switches for controlling the hoist. Fig. 250 shows a front view of the apparatus.

The bevel gears (J) have a ratio of about two to one. Any convenient size may be used, the writer using 4-inch and 2-inch gears taken from an old churning machine. The small gear is arranged on a shaft with a handle, Fig. 256, to turn the large gear. The shaft detailed in Fig. 255 is of iron, steel or brass, and is shown ⅜ inch in diameter, although, of course, any other size may be used to fit the holes in the gears. If such is the case, the holes shown in the detail of the angle bearing (C), Fig. 253, and bearing plate (D), Fig. 254, will vary accordingly. A bearing plate similar to (D) Fig. 254 is used to hold the shaft extending from the mast in the bottom of the box.

The mast is held rigidly upright by means of the legs (A). These are detailed in Fig. 251. They are slotted as shown and soldered to a plate which allows the mast to turn by means of a wood screw screwed through the plate into the top wooden block of the mast. The legs are threaded at one end and bolted to feet, Fig. 257, on the base.

No details of the pulleys will be given, as the average experimenter generally has several around his shop which are adaptable for this purpose. The shape and arrangements of these pulleys, however, are seen in the general view, Fig. 232. For the cables between the pulleys and winding drum, stout string is used.

The double drum electric hoist is mounted on the base and one drum is connected to the boom pulley and the other to the hook pulley. The switches used for reversing the motors of the hoist are of the pole changing type with a dead point in the center. The arms of these are extended through slits in the sides of the box and handles fastened to these extensions. This is clearly illustrated in Fig. 233. Double pole, double throw switches may also be used to control the motors, the switches being mounted on a separate switchboard.

If the reader does not care to have the operation electric, he can construct hand-operated winches for the derrick.

The whole apparatus should be painted a dull black, which gives it a commercial and serviceable appearance.

CHAPTER XXV

The design of the hoist—Calculation of the power needed—Construction of the machine.

Every model maker delights in making models of machinery, so that when they are constructed they have the appearance of the larger article and their limits of usefulness are just as great. The double drum electric hoist, about to be described, comes up to these requirements. Its range of utility is varied. It may be used on small cranes or derricks, elevators, double hoists, traveling hoists, and probably many other uses will suggest themselves to the constructor.

By consulting the plan, Fig. 262, it will be seen that each winding drum is driven by a separate motor, which simplifies the control of the hoists. Eight gears will be required, two of each of the following: 96 teeth, 12 teeth, 48 teeth and 20 teeth. The dimensions between the shafts are for gears of 32 pitch. In case the reader has some other size gears on hand he should, of course, use them, although the distance between the shafts will not be the same. Quite a little power is developed at the winding drum, the apparatus easily lifting several dry batteries. This will be readily seen, as the speed is

$$\frac{12 \times 20}{49 \times 96} = \frac{5}{96}$$ of that of the driving motor, and as the

lifting power varies inversely with the speed the lifting

power is $\frac{96}{5}$ or $19\frac{1}{5}$ times that at the motor.

The two sides (A), Fig. 263, are made of sheet iron or brass about ⅟₁₆ inch thick. The figure shows the location of the holes for the shafts, guy rods and base angles. The most convenient way to bore these holes accurately is to first locate and bore the ⅛-inch holes in each corner of each piece and then fasten the two pieces together with 6-32 screws through these holes and

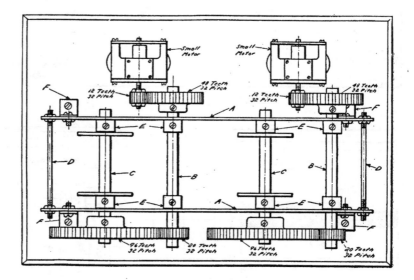

Fig. 262—Plan of the electric hoist

finish the plates to size. The shaft and base angle holes are then located and bored.

Fig. 265 represents the guy rods, four in number, which are ⅛-inch round brass rods and threaded ¾ inch on each end with a 6-32 die. These rods hold the sides in place, as shown in Fig. 262.

Fig. 264 is a drawing of the shaft (B), two of which are needed. They are preferably steel rod, although brass or iron will serve the purpose equally well.

The winding drum (C), Fig. 266, consists of a shaft

on to which are soldered two discs. The best way to
solder these on is to bore the hole for the shaft through
the disc a trifle scant and then force them on the shaft,
after which it should be soldered by heating it with a
blow-torch and the acid flux and solder applied with a

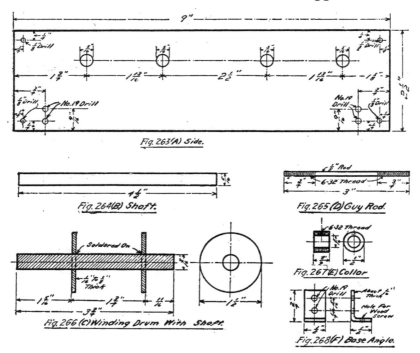

Figs. 263-268—Details of the double drum electric hoist

thin rod or wire and the solder worked in the joint. This
joint should be wiped while still hot and a neat job
should result.

Each shaft requires two safety collars (E), Fig. 267.
They are made of ½-inch brass rods which is bored ⁵⁄₁₆
inch. A 6-32 hole is tapped through the walls of this
piece. Eight, the number of collars required, 6-32 x ³⁄₁₆-
inch round head brass screws will be required for these
collars.

The base angles (F), Fig. 268, are merely strips of metal shaped as indicated. Their purpose is to hold the frame of the hoist to the base. Four are required. Eight 8-32 x 3/8-inch brass screws with nuts will be needed to fasten the angles to the sides.

The hoist may be mounted on a base with the motors or may be fastened permanently with the apparatus it is

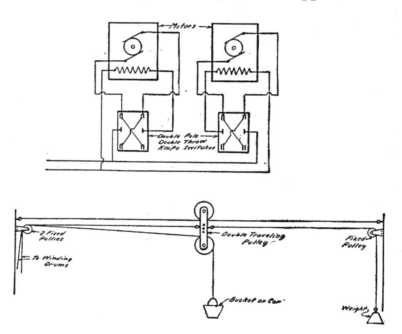

Fig. 269—(Above). Wiring diagram of the electric hoist
Fig. 270—(Below). Arrangement of the traveling hoist

working, as on the base of a derrick or frame of an elevator. If mounted separately with motors, as stated above, a base of oak, 7 x 10 x 3/4 inch, neatly finished with beveled edges, will suffice.

The method of holding the driving gears to the shaft of the motor is by removing the pulley from the motor and threading the shaft with the most convenient size and then slipping the gear on the shaft and locknutting it on. If there is a small hole in the gear it may be driven on the shaft to a forced fit, which is better than the previous method. Almost any type of motor is suitable as long as it reverses easily and quickly.

The wiring diagram is represented in Fig. 269. Double pole, double throw switches are ideal for reversing, although a pole-changing switch may be used if it has a dead point, so that the motor can be stopped without reversing the motor.

A sketch of one of the many uses to which the hoist may be put is represented by the traveling hoist in Fig. 270. By its use the bucket or car may be carried to any place at any height along the cableway. The arrangement is frequently seen where excavating is being done. The details of the pulleys are not shown, same being left to the ingenuity of the reader. At first it will be difficult to stop the car in the desired place, but after some experience, one becomes quite skilful in its operation.

CHAPTER XXVI

A MODEL GASOLENE ENGINE

Machining the castings—Making the crankshaft—The carburetor—Spark plug designs—Ignition.

The engine has a 1-inch bore and 1-inch stroke. It is designed to develop one-eighth horse power but has never been put under brake test. The weight of the engine without flywheel is slightly less than a pound. It is sufficiently light and powerful to drive a large model airplane or a fast model boat.

Simplicity marks the design in every particular. It is of the two-cycle type and has no valves, but merely a by-pass for the exchange of exhaust and fresh gases. The carburetor is of the simple mixing valve type which may be used with gasolene or even house (illuminating) gas merely by an adjustment of the fuel and the air supply.

The motor is air cooled and for ordinary runs no fan has been found necessary. It is designed primarily for portable use in some situation where the movement of the boat or airplane creates sufficient draft to cool the cylinder. For stationary use, part of the radiating flanges might well be turned off and a simple water jacket adapted to the cylinder. This could readily be done by fitting a piece of aluminum tubing over two of the flanges, which would, of course, be turned down to make a tight force fit, the intervening flanges being removed.

The castings throughout, with the exception of the piston, are of magnolite, an alloy having a remarkable tensile strength with the light weight of aluminum. This

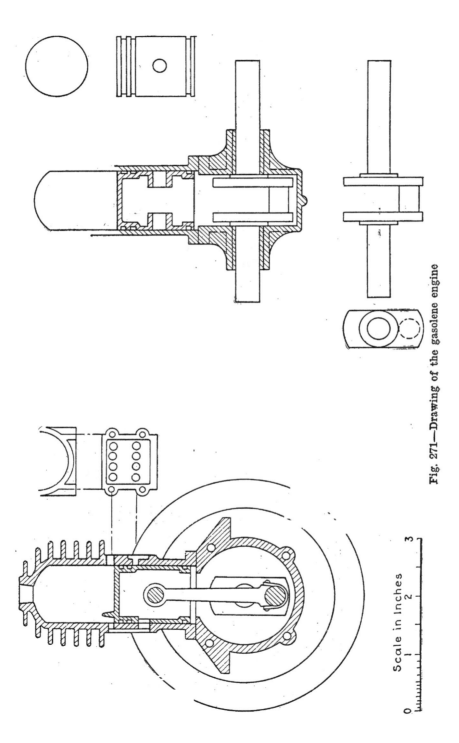

Fig. 271—Drawing of the gasolene engine

Scale in Inches

0   1   2   3

metal machines very nicely and the castings come from the foundry in a remarkably clean condition. The natural color is that of aluminum, and the model, therefore, presents a fine appearance with but little cleaning up.

The cylinder casting is bored, of course, and very little stock need be taken out. The casting may well be held by the radiating flanges in a three-jaw chuck for the boring operation. Care should be taken to see that the chuck jaws have a good, firm bearing at all points be-

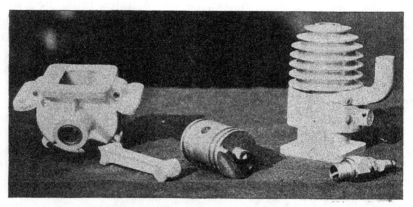

Fig. 272—The main parts of the engine machined

fore attempting the cut. The casting is so clean and accurate, however, that ordinary care will insure this.

The best method is, of course, to bore in the usual way with several light cuts, and to finish with a parallel reamer. This gives as nearly perfect a job as is possible with the tools at the disposal of the average model maker.

The next step is to face the end of the cylinder casting at the place where it is to join the crankcase. This operation will have to be done cautiously in order to prevent a "bite." The casting may then be removed from the chuck and mounted upon a wooden arbor with the dead center either passing through the spark plug hole or, preferably, engaging a center hole in a plug screwed into the spark plug opening.

The turning of the flanges is done while the casting is on the arbor. A special tool may be ground for this purpose from a piece of self-hardening steel. The contour is well shown in the photographs, while the dimensions may be taken directly from the sectional drawing, Fig. 271.

The ports and by-pass holes should not be drilled until the engine is assembled, as these holes must occupy certain very important positions in relation to the travel

Fig. 273—The complete engine assembled, with ignition apparatus

of the piston. Aside from filing the port and by-pass cover seats, and drilling the holes in the cylinder base for holding down screws, the work on the cylinder casting may rest at this point for the present.

The two halves of the crankcase are cast from the same pattern. The casting may be held in the three-jaw chuck with the jaws gripping the bearing hub. The first operation is to face off the end of the casting. Then may come the turning of the interior. This should be finished throughout in order to permit the crank to clear by a small margin. The hole for the bearing bushing may then be bored. The bushing is of brass, turned from round stock and forced into the magnolite crankcase casting. When both halves have been faced and bored and the bearings fitted, the castings may be assembled on a piece of steel rod passed through the bushings. With

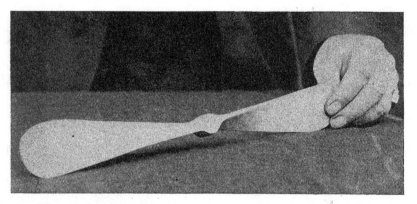

Fig. 274—Aluminum aërial propeller constructed for use with the engine

the halves thus lined up, the screw holes may be drilled and tapped and screws inserted.

The holes in the bushings should have been drilled or bored ¼₄ inch under size, and, after the crankcase has been assembled, a ⅜-inch reamer should be run through the two bushings to insure correct size and absolute alignment. Before taking the crankcase apart after this finishing operation, the two halves should be filed up at the joint so that they may be placed together in exactly the same position after the crankshaft has been introduced.

The piston is of cast iron and it is packed with three rings. The first operation in machining is to grip the casting in the chuck, taking a cut out of the open end, and facing off this end. The casting is then removed from the chuck and mounted upon a steel arbor with the tailstock center engaging a center hole bored in the small

Fig. 275—A close-up view of the timing device and carburetor

projection that may be noticed in the photograph showing the engine dissected.

The outside of the piston may then be turned, the finishing cut being taken with a very smooth, round nose tool, the feed being slow and the cut very light. The speed should, of course, be high in this case. The final "working in" finish may be taken with emery cloth and oil, the former being of the finest grade. The fit of the piston need not be absolute, but it should be so good that the piston will sink very slowly to the bottom of the cylinder against the air cushion before the packing rings are in. The grooves for the rings are turned with a parting tool ground square and to the right width.

The designers attempted to use cast iron piston rings, but found that they could not spring this very small diameter of ring over the piston without cracking the ring. They then turned to cold rolled steel and found rings of this material perfectly satisfactory. The stock was bored-out to the correct internal diameter, which is just a trifle larger than that of the ring channel in the piston. Pieces of exactly the correct width were then

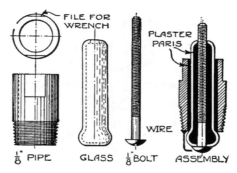

FILE FOR WRENCH

PLASTER PARIS

WIRE

$\frac{1}{8}$" PIPE   GLASS $\frac{1}{8}$" BOLT   ASSEMBLY

Fig. 276—Details of a simple spark plug for the engine

parted from the stock and each piece or ring split diagonally in the usual manner. The split rings were then compressed, one at a time, of course, and gripped on a nut arbor constructed for the purpose. Then the outside cut was taken on the compressed ring to make its diameter a sliding fit into the cylinder bore. This resulted in perfectly round packing rings that were sprung over the piston and into the grooves in the orthodox manner.

The connecting rod is a magnolite casting drilled at the small end for the wrist pin which passes through the piston and filed to bear upon the crankpin at the large end. A brass strap passes under the crankpin, holding the latter to the connecting rod bearing. This method is perfectly satisfactory, as all of the work done by the connecting rod is on the down stroke. The strap serves merely as a means of holding the connecting rod and crankpin together.

The crankshaft may be a forging, steel casting, or a built-up job, or it may be worked out from the solid if

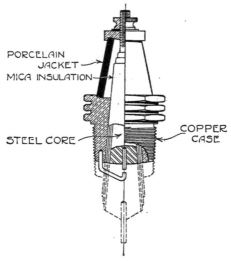

PORCELAIN
JACKET
MICA INSULATION

STEEL CORE

COPPER
CASE

Fig. 277—Showing how a standard market spark plug is reconstructed for use with the gasolene engine

the builder has the patience. The built-up crankshaft is perhaps the easiest if it is well and properly made.

The carburetor is a simple mixing valve in which either gas or gasolene may be admitted in varying quantities by means of the two adjusting screws to be seen in the illustrations. The fuel valve is of the needle type and its opening breaks into the seat of the poppet air valve. The latter is limited in its travel by means of the adjusting screw in the top of the chamber.

The prospective builder is advised to refer to any two-cycle engine catalogue wherein such a mixing valve will be found illustrated in section, providing the engine is one of the low priced marine motors on which no carburetor is used. The device used on this little engine is a replica of the mixing valve referred to, only it is perhaps less than half the size of the standard ones.

The timing device for the ignition circuit is simplicity itself, but effective withal. It consists merely of a fiber ring in which a flush steel setscrew forms the "ground" contact. As the ring rotates the setscrew makes instantaneous contact with the radial brush in the tubular holder held by means of an insulating bushing in the timer arm. The spark may thus be advanced or retarded as the occasion may demand merely by moving the timer arm.

In an early paragraph we referred to the importance of having the engine assembled before the ports were drilled. With reference to the sectional drawing, the reader will note that the by-pass port and the exhaust port are exactly in line and open when the piston is at the extreme limit of its downward travel. It will also be noticed that the exhaust port is slightly larger than the upper by-pass port. That is, they are uncovered by the piston at this point. The lower port of the by-pass is uncovered by the hole in the piston at this point to permit the compressed charge in the crankcase to pass through the by-pass and into the cylinder against the baffle plate. This latter hole is drilled in cylinder and piston at the same time to insure alignment. All of these holes are *open* at the *extreme downward point.*

As the piston moves upward, compressing the charge in the cylinder, all ports are covered by the piston. A vacuum is created by the upward stroke in the crankcase so that at the extreme upward limit of the piston's travel a charge of fresh gas will be drawn into the crankcase. In order to permit of this, the inlet port must be drilled just below the piston when the latter is all the way up in the cylinder. These relations are of the utmost importance to the efficient working of the engine.

A cast iron flywheel is fitted to the model. It can readily be displaced by an aluminum water or aërial propeller. A partly finished 14-inch aërial blade is shown in Fig. 274.

The engine is oiled by mixing ½ pint of good automobile cylinder oil with every gallon of gasoline when this fuel is used. This mixture is drawn into the crankcase and the cylinder on the suction stroke. The oil enters as a finely divided spray which does its work most effectively. If house gas is used, the oil will have to be put into the crankcase where it will splash on the downward stroke of the piston. The oil level must be kept low in this case, as otherwise too great a quantity of oil will be drawn into the cylinder through the by-pass and cause carbonization, smoky exhaust, and poor combustion.

The model builder generally finds it difficult to purchase small spark plugs for model gasolene motors. The spark plug shown in the illustration can be made in the shop in a very few minutes and will hold up under service for some time.

The case of the plug is made from a piece of ⅛-inch pipe bored out so that a piece of 3⁄16 glass tubing may be placed in it. The pipe is filed flat, as shown, so that it may be put in place with a wrench. The glass tube is first heated and bent into the shape depicted in the sketch. This is to prevent it from being forced out of the pipe by the pressure developed in the cylinder of the gas engine. A slender brass machine screw is arranged concentrically within the glass tube. The space between the glass tube and the wall of the pipe and the space between the machine screw and the glass tube is filled with plaster of Paris, which, when thoroughly dried out, forms a very good insulating compound. A tiny hole is drilled on the lower edge of the pipe and a small piece of wire is inserted in this and soldered in place. This is arranged in a small loop and the electrical discharge of the coil takes place between this and the head of the brass machine screw. Another plug suitable for this engine and which can be easily made is shown in Fig. 277.

To make this plug, procure the core or inner member of any of the *two-piece tapered core, mica* plugs, such as Splitdorf, Benton, Wright or Mosler. Saw off the projecting mica and finish off the remainder with a file to a smooth rounded surface, as shown in the drawing. Then insert the wires in holes drilled for that purpose, either by shrinking the core over the wire or clinching the wire in with a punch. The wires may be arranged in a variety of ways, many of which will suggest themselves to the builder. The easiest and simplest would be to fasten the wire in the center electrode and bend it so that the spark would jump from the wire directly to the shell.

Cooling vanes may be cut in the hexagonal portion of the shell, if there is enough material, but with a mica plug this is unnecessary.

The only disadvantage is that a large plug must be purchased to start with; however, the finished plug is well worth it, being every bit as good as the plug from which it was made and easier to clean.

The spark coil for the engine can be purchased in the open market or it may be made. A small $\frac{1}{8}$-inch spark coil would be sufficient in case the coil is purchased. If the coil is made by the builder, it should be so designed that it will give a small, heavy spark about $\frac{1}{16}$ inch long.

CHAPTER XXVII

A MODEL ELECTRIC LOCOMOTIVE

The prototype—Construction of the locomotive—Its driving motors—
Finishing the locomotive.

This interesting locomotive opens a new field of remarkable fascination to those interested in railway models. Its prototype is the giant duplex electric engine of the Pennsylvania Railroad, which is of especial in-

Fig. 278—The complete electric locomotive of the Pennsylvania type

terest for its unique system of power transmission to the wheels. These engines comprise two identical permanently coupled sections having two pairs of drivers and a four-wheeled bogie truck each, and carrying a single motor, which is mounted with its shaft transversely of the

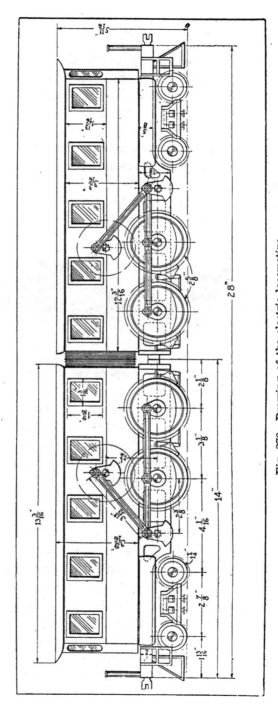

Fig. 279—Drawing of the electric locomotive

frame, in each unit. From a crank disc on each end of the motor shaft, angularly placed connecting rods lead to a crank mounted on a jackshaft, which in turn is coupled to the drivers by side rods of the type employed on steam machines. This construction, unusual in electric locomotives, makes the model a very attractive one. The model, which exhibits exquisite workmanship in its construction, measures twenty-eight inches long over all.

Fig. 279 gives complete dimensions and details of the engine units, the two units being, of course, identical in

Fig. 280—The locomotive chassis, showing the driving motors and the method of gearing them to the drivers

every particular. The main frames are cast of brass or bronze, in one piece with the axle boxes, jackshaft bearings, and bearing blocks for the shaft corresponding to the armature of the driving motor in the prototype. No difficulty is likely to be experienced in finishing these parts, aside from the necessity of obtaining perfect parallelism of all the shaft bearings, in order to insure the easy working of the connecting rods. This method of construction greatly simplifies the model, and while it provides no springing whatsoever, attempts to employ spring suspension have never proven particularly successful in a model of this size, since it unduly stiffens the engine, and springs are undoubtedly best dispensed with.

The wheels are of cast iron and are similar to those employed in steam-driven designs. The jackshaft cranks should be carefully counterbalanced. In quartering the crankpins of these members the same method was used as was employed in the construction of the Pacific type of steam locomotive described in Chapter XXVIII, which greatly simplified this work. The wheels were first mounted on their axles in approximately their correct positions and then a U-shaped jig is placed over the wheels and the crankpin holes drilled exactly 90 degrees apart, insuring that the cranks will be in their correct relation when the side rods are fitted. The bogies are simple four wheel trucks very similar to that of the Pacific engine, and need no further explanation. The side rods can be cast of brass or gunmetal or worked out from the solid piece, and present no particular difficulties. The brake rigging may be simply dummy or actually operative, and small solenoids may even be fitted to give electro-magnetic brake control of the engine. A small bolt with coil springs as cushions is used to couple the two sections.

The cabs are made of sheet brass, well braced and soldered, and spring over the platform of the base. They carry the searchlights, which may be fitted with small flashlamp bulbs, small dummy pantographs, which, in the large engines, are used in yard switching where overhead contact is necessary, the bells, and ventilating hoods. The windows may be either of glass or mica. The finish is in black and gold, identical with the large Pennsylvania No. 30, and the lettering is easily done with small paper letters pasted on the cab. Fitting the cowcatchers and end platform railings completes the model.

A number of arrangements of driving motors were experimented with by the builder, starting with tripolar permanent field motors driving crown gears with a reduction of five to one, the object being to reverse the engine by changing the polarity of the exterior track connections. The gear reduction proved insufficient, however, and a worm and pinion combination giving a forty to one reduction was attempted. This arrangement may be seen in the photograph of the engine with the cabs removed. While possessing great power, the reduction was here too great, and the proper solution to the problem seems to consist either of a combination of worm and spur gearing to give a higher speed, or, if one can be secured, a double thread worm and forty tooth

Fig. 281—Side view of the electric locomotive

pinion giving a twenty to one reduction. The permanent magnet motors were not an unqualified success, and it is undoubtedly advisable to abandon the directly reversible feature and employ standard motors of either 110 volt or battery types. There is ample room in the cab for any arrangement that may be desired. The current collecting mechanism, of course, depends upon the disposition of the third rail.

CHAPTER XXVIII

A MODEL STEAM LOCOMOTIVE

The locomotive's prototype—General design—Construction—Fittings.

The model described is of a Delaware, Lackawanna & Western, Pacific type, passenger locomotive, built by the Lackawanna Railroad about June, 1913. At that time it was the largest passenger engine of its type in the world, the engine alone weighing 284,000 lbs. and the

Fig. 282—The Lackawanna locomotive

tender 165,700 pounds in working order, and having a capacity of 14 tons of coal and 8,000 gallons of water.

The construction of the model is started with the main frames, which were made of ⅜₆-inch cold rolled steel, drilled, filed and sawed out to represent the usual bar frame of American practice. Although this frame of ⅜₆-inch thickness seems excessive, it furnishes an excellent foundation for the valve gear frame and cylinder saddle. This heavy framing only extends about an inch behind the rear driving wheel axle box. From here to the end of the frame is only ⅛ inch in thickness, the rear end being tied by a brass casting and the forward end by the cylinder castings which embrace the frame, as shown in front view of engine.

The next step to be taken is the axle boxes; these are of brass and have ample bearing surfaces, being no less than ⁹⁄₁₆ inch wide. The compensating spring hangers and brackets are built on these axle boxes, as shown in Fig. 284.

The trailing truck comes next. This has an independent framing of ⅛-inch cold rolled steel, connecting

Fig. 283.—The locomotive mounted upon its testing stand

cast axle boxes at each side and pivoted just in back of rear driver. The weight is carried and sprung on a sort of U-shaped hinge. The spring is held at the rear end by a bracket bolted to the main frame and the forward end by a hanger connected to an equalizing lever anchored to the main frame by pin. The forward truck is the ordinary equalizing lever type, which needs no further explanation.

The wheels are all of cast iron, the drivers being 2⅝ inches in diameter, trailing wheels 1¹³⁄₁₆ inches diameter, and front truck wheels 1³⁄₁₆ inches diameter. In quartering the driving wheels for the crankpins, a novel method

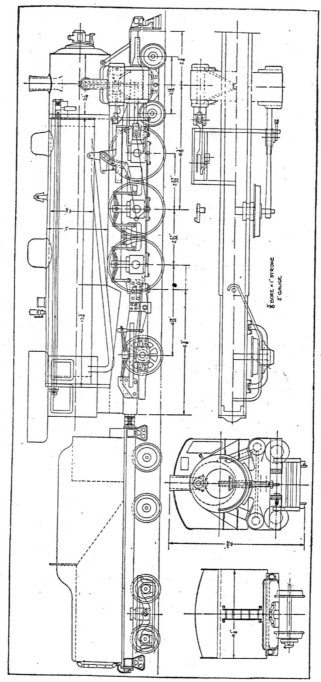

Fig. 284—General details of the Lackawanna locomotive

was used which proved very successful. The wheels were
first taken in hand, bolted to the face plate, rough turned,
centered and drilled ¼ inch for axle. The axles were
turned between centers out of large enough material to
provide for turning all over to ⅜ inch diameter. Then
the shoulders were turned to ¼ inch; a driving fit for
driver-wheels which are now driven on to axles with the
crankpin bases as near 90 degrees apart as possible.
To drill the holes for the crankpins, a jig was made, as
shown by the drawing; and when the holes are drilled
and pins fitted, it will be found that there is nothing to
be done in the way of fitting side rods, as everything will
work out absolutely corerct. The crankpin for the main
driving wheel is extended beyond the driving rod for the
return crank of the valve motion.

The cylinders are cast of gunmetal, as in the original
engine, in halves, and bolted together at the center.
Steam is distributed by a piston valve with inside admission. The piston valve has no exhaust lap and very little
lead is given to steam admission. The piston valve is
first ground in and then grooves turned in and packed
with a graphite steam packing which wears very well and
furnishes lubrication for the valve chamber. This packing was found necessary owing to the fact that when the
engine was put under steam for the first time, the steam
would blow through the valve as soon as lubrication was
gone, in spite of the fact that air tanks at the side are
displacement lubricators. The piston valve rod has a
guide and crosshead, as shown on plan and elevation.
Crosshead and crosshead guides are machined from cold
rolled steel, and the guides are held in position by a
bracket fastened to the valve link frame, making a very
rigid construction. The method of fixing crosshead
guides is always a vexing problem. These guides are
threaded at one end and screwed into the back cylinder
head, then held apart the proper distance by a bracket

which is made from one piece of cold rolled steel, and suspended from the valve link frame. The cylinders are also provided with release valves. These are ball valves and are released by moving the release rod backward or forward, the valves working on a taper.

The valve motion is the full Walchaerts valve gearing, brought down absolutely to scale and actuating the valve perfectly. Eccentric rod, valve rod, combination lever and crosshead union link are all made of German silver; valve rods and piston rods are also of the same material. The valve link is made of cold rolled steel, case hardened. Reversing of the engine is done by means of the bell crank lifting valve rod up or down in the link, as the case may be, and is operated by screw reverse in cab.

The boiler is built on the well-known and tried Smithies principle. The boiler proper is built of copper tubing $2\frac{1}{8}$ inches diameter and $5\frac{1}{4}$-inch copper water tubes put into the boiler, as shown in the elevation. The back and front plates are of cast gunmetal. The front plate also has a throttle valve casing cast on it and is of the sliding type, working on the inside of the steam collecting tube. Steam is supplied to the cylinders by separate steam pipes. Each steam pipe leaves the valve casting, goes to the back of boiler, to firebox, returning to smoke box and then to cylinder by way of outside steam pipes. The boiler shell is built of Russian iron plate and well lagged on the inside with sheet asbestos. Both domes are dummies. The boiler fittings comprise water gauge, pressure gauge, check valve, blower valve, whistle bell, safety valve and Westinghouse air pump on left side. The blower is quite a novel idea which was gained from Mr. Wardlaw, the builder of the prize-winning fire engine at the last Model Engineer Exhibition and pictured in the frontispiece. Ordinarily, it is difficult to get the blower to blow true to the center of stack, but here it is quite a simple matter. The blower pipe is brought along underneath boiler and connected by union to the blower nozzle at the base of the exhaust pipe, which at one operation brings blower in true line with the stack. Reference to the drawing will bring this to view more plainly.

CHAPTER XXIX

Many model makers have probably often wished they could construct a small gyroscope railroad, but owing to the scarcity of literature on the subject of practical gyroscopic balancing, they hesitated to start work upon the project. The writer has spent considerable time upon this particular study and has developed and constructed a small monorail car which maintains almost perfect equilibrium and presents few difficulties in building. Several original features are incorporated in the design of the gyroscope used.

Although a gyroscope completely obeys the laws of gravity, its action is very mysterious to the layman and a few words on the subject will probably be welcomed before the actual construction of the car is taken up. Reference is made to Figs. 285 and 286, which will assist in explaining some of the principles involved. A gyroscope may be described as a rapidly revolving wheel which resists any tendency to change its plane of rotation. In Fig. 286 the gyroscope is mounted so that the axis of rotation (A B) is vertical, while the axis of precession is transversal to the line of support (C D). As long as the gyroscope remains in a perfectly vertical position, its plane of rotation will continue to be vertical. If it is tilted as shown in Fig. 285, however, the gyroscope will have a noticeable tendency to dip, and this is called "precession." This peculiar action of the gyroscope can be used to advantage in balancing a car by means of a "precession fork." The idea of the precession fork is to enforce increased precession upon the axis A B the instant the car does not stand vertical upon the rail. Increasing the precession in the direction in which the gy-

Fig. 285—Showing the procession of a gyroscope in the direction indicated by the arrows

roscope tends to precess will create a new torque in the direction of the vertical and at the same time the axis A B will turn back to a vertical position. If the forced precession, which the gyroscope actually stores up within itself, is not sufficient, the car will tip over, and if this

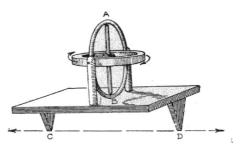

Fig. 286—The gyroscope mounted so that its axis of rotation is vertical

precession is too great, the car will right itself too quickly and fall to the other side before the gyroscope returns to the position necessary to maintain equilibrium. Friction is the factor which produces the forced precession, therefore the coefficient of friction must not remain constant and the proper metal to use on the precession fork must be found.

The general constructional details of the car and gyroscope will now be treated. The body and base of the car is made of wood, although the base can be made of light metal if the builder so desires. The top is made of tin and is provided with small ventilators. A small

Fig. 287—The model gyro-car balancing itself upon a wire

headlight is also arranged on the front, and this need not only be ornamental, but can be supplied with current from the source that drives the car. The windows of the car can be made from glass or thin mica, which is easier to cut and looks just as good as glass. The wheels, of course, are made with a double flange, as they will slide off a single rail if made in the usual manner. The trucks are shown in detail in Fig. 290 and the reader is cautioned to mount them exactly in the center of the platform. Otherwise, the car will not be balanced properly and the gyroscope will be called upon to work harder than would otherwise be necessary.

Especial attention is drawn to the method of mounting the motor. The necessity of making the driving truck a compact unit will be readily seen, as this must be done so that the car will be able to turn curves. If the motor was mounted on the floor of the car, it would be a very difficult matter to arrange the driving mechanism so that the truck would be flexible enough to turn curves. If it is mounted in the manner shown, no trouble will be experienced. The worm gear reduction will depend largely upon the speed of the motor. The entire frame of the truck can be bent into shape from heavy gauge brass and either riveted or soldered. The cradle for the motor will have to be placed a little to one side, and while great effort should be made to balance the car by placing everything on board in a central position, an exception will have to be made in this case. The car can

Fig. 288—The model gyro-car balancing itself upon a single rail

be counterbalanced with small lead weights after it is completed. The other truck of the car is made exactly like the driving truck, with the exception that no provision is made for the motor.

The arrangement of the gyroscopic apparatus is clearly shown in the sketch. The gyroscope is driven with a motor which has a speed of at least 4,000 R.P.M. The motor shaft is connected directly to the gyroscope and both the gyroscope and motor are mounted in a gimbal so that the gyroscope is free to move in any direction parallel to the motion of the car; that is, either backward or forward. The axle of the gyroscope is placed between the precession forks and these are cut from sheet brass. They are mounted on two small posts. The clearance between the axle of the gyroscope and the precession fork should be very small, and yet the gyroscope should in no way be hindered from revolving

Fig. 289—The motor used to spin the gyroscope

with absolute freedom. The gyroscope and all its parts should be well lubricated with a very thin machine oil.

Owing to the fact that small gyroscopes cannot balance a weight of over two pounds, the possibility of driving a car two feet long with dry batteries becomes very remote. By arranging an improvised trolley on this model, quite a heavy current can be carried to the motors without any addition in weight. To do this, a wire is suspended directly over the rail or wire upon which the car is to run. Each wire is connected to the source of current and the current is led into the car by a small cord which runs to a little pulley or trolley that runs along the copper wire. An ordinary trolley cannot be used, as this would tend to prevent the car from tipping and the gyroscopic effect would be lost. The small flexi-

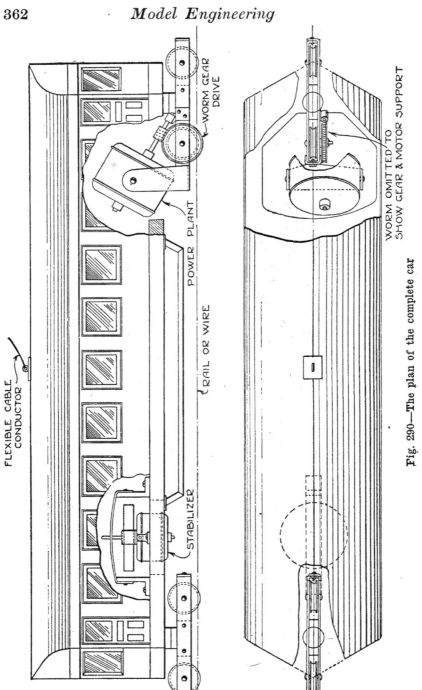

FLEXIBLE CABLE CONDUCTOR

WORM GEAR DRIVE

POWER PLANT

RAIL OR WIRE

STABILIZER

WORM OMITTED TO SHOW GEAR & MOTOR SUPPORT

Fig. 290—The plan of the complete car

ble cord permits the gyroscope to do its work, and in this manner unlimited power may be provided for the motors. If it is desired to run the car on a rail in place of a suspended wire, a method of making rails from

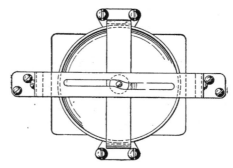

Fig. 291—Top of the gyroscope, showing the precession fork

heavy wire and mounting them on ties is shown in the drawing. In making cars of this type, the following facts should be kept in mind by the builder:

The center of gravity should be kept as low as possible.

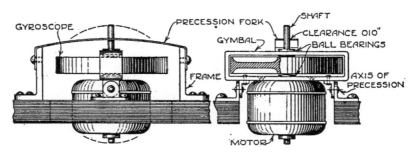

Fig. 292—Details of the gyroscope

The car should be perfectly balanced.
The weight should be reduced to a minimum.
If these instructions are carefully carried out the

builder should have no trouble in producing a successful gyroscope model.

If the car is run on a rail in place of a wire, the trolley wire will be suspended in a different manner and a different type of trolley will have to be used also. This is necessary because the trolley must be supported by poles and the trolley used on the single wire cannot get

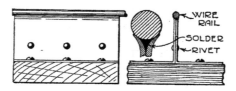

Fig. 293—How rail joints are made, using heavy wire as the rail

past the wire supports. In this case, an "L" shaped trolley can be employed so that the trolley wire supports can be fastened to one side of the trolley wire.

By use of the heavy wire for the rail, a large railroad can be built with very little cost. A rheostat can be used in the motor circuits so the builder can control the speed of the car at will.

CHAPTER XXX

A MODEL CATERPILLAR TANK

Framework of the tank—Tractors—Driving motors and gearing—Reel for feed cable—Covering the tank.

The construction of a model tank that limbers along under its own power just as the big fellows do offers a very interesting project for the model maker.

The framework of the tank is made almost completely of wood. The construction is started by cutting out the two side pieces from ½-inch stock. At the widest point these pieces measure 9 inches and they are 25 inches overall in length. The shape of these sides can be seen by referring to Fig. 294, which shows the side of the tank. The drawing is not dimensioned, as it is believed most builders would alter the design to suit their own tastes and that the general directions would suffice. If the builder desires to make the tank exactly as shown, however, he should not experience any trouble, as the various parts can be proportioned in accordance with the dimensions given for the two side pieces. The proper size of the various parts can also be seen from the drawings, as they have been made exactly in proportion.

After the side pieces have been cut and trimmed up with a wood rasp, several cleats are nailed to their inner sides. The shape and position of these cleats is shown very clearly by the dotted lines in the drawing of the side of the tank. These cleats serve two purposes. They are used to nail the crosspieces to and they are also used to offer better bearing support to the wooden rollers that carry the belt. The crosspieces are nailed in place as shown. A large board is nailed to the bottom of the two cleats that run along the lower end of the tank from the big driving pulley in the back to the large pulley located near the front at the bending point of the belt. This

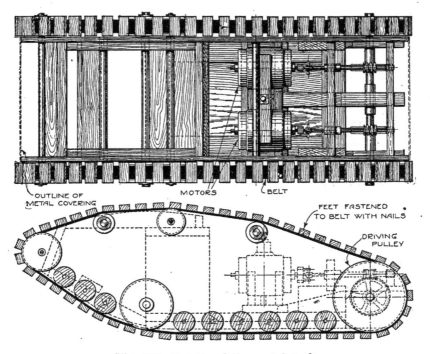

Fig. 294—Details of the model tank

large board acts as a floor and support. The other crosspieces act merely as braces.

The rollers are made next and several different sizes are needed, as will be seen. Twenty of the small rollers will be used and these should not be under one inch in length—their length should be just equal to the width of the belt used. The pulley or rollers measure 1½ inches in diameter and they may be cut or sawed from a single piece of the proper size. If the builder does not have access to a wood-turning lathe, the rollers may be obtained from a planing mill. The bigger rollers are provided with flanges which keep the belt in the proper position, the smaller rollers merely acting as supports. The small rollers are held in place by ¼-inch cold roll rods provided with a thread at one end and a hole at the other through which a cotter pin passes to prevent the roller from slipping off. The threaded end of the rod is turned down so that there will be a shoulder. A

Fig. 295—The model tank with the sides removed

small washer is placed over the threaded end of the rod, and this comes to rest against the shoulder. The rod is then placed in the hole in the wooden side and a nut screwed on the opposite side. This holds the rod rigidly in place and the roller can then be put in position and locked there by means of the cotter pin. It is understood, of course, that the rollers should be permitted to revolve freely and with a minimum of friction. The large rollers are not held in this manner. In place of the small rod that acts as a bearing for the smaller rollers, a rod runs from one side of the tank to the other, protruding at each side. These rods should not be smaller than ⅜ inch, and, with the exception of the main driving roller on the back of the tank, they are held in place by means of a cotter pin and washer.

The tractors of the machine are made from two lengths of one-inch leather belt. Small wooden blocks are fastened to the belts by means of nails or brads. The blocks can be nailed close together or spaced a short distance apart. If desired, the nails can be of such a length that they will protrude a short distance on the opposite sides of the blocks so the tractors will obtain a better grip on the surface the tank is traveling on.

The driving units will now be considered. The driving rods are provided with large gear wheels which are driven by worms on the motor shafts. It will be seen that each of the large driving rollers are provided with a separate shaft and the ends of these are both supported

Fig. 296—The rear of the tank with the cover removed to show the mounting of the driving motors

by the one-inch wooden strip that runs from the motors to the crosspiece on the back of the tank. In order to steer the tank and to avoid complications in the driving mechanism, two separate motors are used; one for each tractor. The speed reduction given by the worm drive on the model the writer constructed was 40 to 1. Of course, this will not hold true in every case, as the reduction necessary will depend entirely upon the speed of the motors. The motors are connected to the driving shafts by means of a knuckle or improvised universal joint. This is quite necessary to prevent binding and constant readjustment of the bearings, as the wooden supports are not capable of holding themselves rigidly in place and will therefore cause trouble. This universal

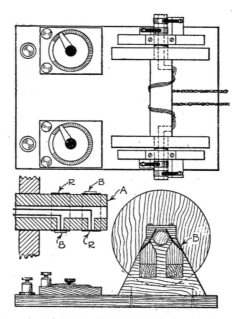

Fig. 297—The reel and commutator device for the feed cables of the model tank

joint is very simple and merely consists of a pin or "tee" on the motor shaft proper running in a U-shaped piece on the driving shaft. The photograph shows this arrangement clearly, as well as the method of mounting the motors in the cradle and clamping them down.

The tank is now ready to be covered with galvanized iron. Two pieces are first cut to cover the sides. These are drilled full of small holes arranged systematically.

Small rivets are then placed in the holes and hammered down. The top is covered in the same manner. If the builder desires, he can arrange the guns on the sides and front. To carry the model nearer to perfection, it can

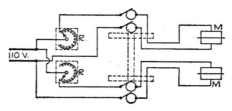

Fig. 298—Wiring diagram of the motors and controlling rheostats

be daubed with various colored paints to represent camouflage.

Being that the driving motors of the tank obtain their power from the lighting circuit, it will be necessary to have the tank pull its feed cables with it. An arrangement devised by the builder is shown in Fig. 297. Two 50-foot flexible cords are wound up on an especially con-

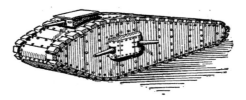

Fig. 299—Drawing showing how the tank appears when completely assembled

structed reel. This reel is provided with four copper rings and brushes so that the current can pass into the cables while the reel is revolving. Mounted on the board with the reel are two small rheostats, one for each motor. By varying the speed of the motors, the tank can be steered nicely. If one motor is stopped, the tank will pivot itself and turn very shortly.

The ingenious builder will readily see the possibility of mounting a small caliber gun in one of the gun turrets and actuating it by means of a solenoid and plunger. The plunger is attached directly to the trigger of the gun. If blank cartridges are used, a very realistic effect can be had. The feed wires for the solenoid can be carried on the same reel that holds the wires for the motors.

CHAPTER XXXI

The barrel of the gun is to be considered first. This measures $13^{11}/_{16}$ inches long over all. It has a bore of $5/_8$ inch and is $2^1/_4$ inches in diameter, at its largest point. The barrel should be turned out from a piece of $2^1/_2$-inch cold rolled steel and is tapered as shown in the drawing. The breech of the gun measures $1^3/_4$ inches outside diameter and this tapers to $1^1/_2$ inches at the center of the gun. From this point it tapers from $1^3/_8$ inches to $^{15}/_{16}$ inch at the muzzle. After the barrel is turned to the proper shape and dimensions it is bored out and reamed. While the original model is rifled, the builder is not advised to do this unless he is especially qualified to accomplish the operation, as it requires great skill and elaborate equipment. The gun, of course, can be fired without being rifled and rifling adds no great value to the finished model.

The hole for the breech mechanism is made next by boring a hole $^{15}/_{16}$ inch, drilling transversely through the barrel of the gun, as shown in the sketch. After this hole is drilled, it is finished square on one side by means of a file. The rear end of the barrel is left open and the hole is slightly beveled on the edges to permit the free entrance of the shell. After the muzzle is turned and drilled, it should be polished with fine carborundum cloth.

The breech mechanism is made next and this is very simple, consisting merely of a solid piece of cold rolled steel cut to the dimensions shown. This is inserted in the transverse hole through the breech of the gun after the shell has been put in place. The handle shown actuates a small eccentric which is turned after the breech is in place and engages with a small groove cut in the inside of the barrel. The method of making this locking

Fig. 300—The model siege gun completely assembled and elevated at 45 degrees

arrangement is shown in the detail drawing. A small plate which covers this mechanism is sawed from a piece of 1-inch cold rolled steel and fixed to the breech with small screws. A small hook, with a thread at one end, is then made and screwed to the breech. This is used to hold the chain, the other end of which is attached to the eye bolt on the lower end of the carriage frame. The handle or wrench which is used on the locking arrangement is cut from cold rolled steel or brass and has a square hole in it which fits over the shaft of the eccentric. Owing to the peculiar shape of the breech, it will be impossible to do any of the work on the lathe. It would not be found difficult, however, to form this by other means, as with careful filing and grinding it can be accurately made. The supporting shafts should now be fitted to the sides. These measure ⅝ inch in diameter and they may be threaded and screwed in place by drilling and tapping a hole in the barrel to receive them.

Although the sighting mechanism is not essential, it adds greatly to the general appearance of the weapon, and it is not difficult to construct. A small, sharp-pointed pin is inserted midway on the barrel between the muzzle and the breech. The other part of the sight is attached to the extreme rear of the barrel and consists of a small scale arranged to slide vertically and provided with a small set screw, by means of which it can be adjusted to various heights. On the top of the scale there is filed a small notch, by means of which it is possible to sight the gun with the pin in the middle of the barrel.

The carriage of the gun will be considered next. This can be made either of cold rolled steel or brass, as the builder desires. The sides of the carriage can be made up nicely from ⅛-inch brass plate and the edges should be bent at right angles and drilled as shown. The two sides of the frame are held together with two cross bars of cold rolled steel which are bolted in place. The end of the frame which rests upon the ground is cut as shown, and covered with sheet brass or steel, bent to the same shape as the end of the frame pieces. A small hole is drilled in the sheet brass which runs across the back of the frame to provide for a chain, which prevents the gun from flinching. If a shaper is at hand, the frame pieces

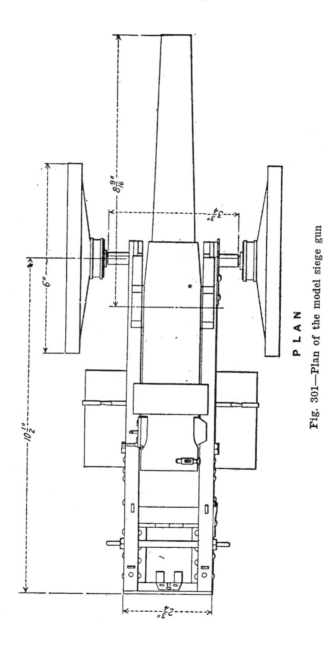

PLAN

Fig. 301—Plan of the model siege gun

can be made from cold rolled steel measuring 1½ inches x ½ inch x 11 inches. Although this will make a very substantial piece of work, there is no objection to the use of brass, as the whole gun is painted black after it is made. Two small folding steps are placed on each side of the gun and attached to these is a small steel rod which drops to a vertical position when the step is raised and thus holds it in place. These steps are used when the gun is being loaded and not when the barrel is being adjusted, as many would think.

The elevating screw, which is used to adjust the range of the gun, is next made and this is actuated by the handles that appear on each side of the carriage. Each handle is attached to two separate shafts, at the opposite end of which is fixed a small beveled gear. These two gears on the small shafts engage with a larger beveled gear that is fixed to the elevating screw. As these handles are turned around, the barrel of the gun can either be raised or lowered, depending upon the direction in which the handles are turned. Although the elevating mechanism on the original model is somewhat elaborate, the builder can alter the design to suit his own desires, as this motion can be very easily produced by a more simple method. One end of the elevating screw is attached to a rod which runs from that point to the middle of the bearing supports, where it is fixed to a rod which runs across the carriage. This rod is free to move with a vertical motion as the gun is raised or lowered.

The bearing supports are cut from cold rolled steel, and unless they are made accurately the barrel of the gun will not move freely. The bearing supports consist of two separate parts. The part upon which the barrel actually rests is bolted to the frame of the carriage and the other part consists merely of a small piece which slips over the lower piece of the support and prevents the shaft from slipping out of place. This is quite

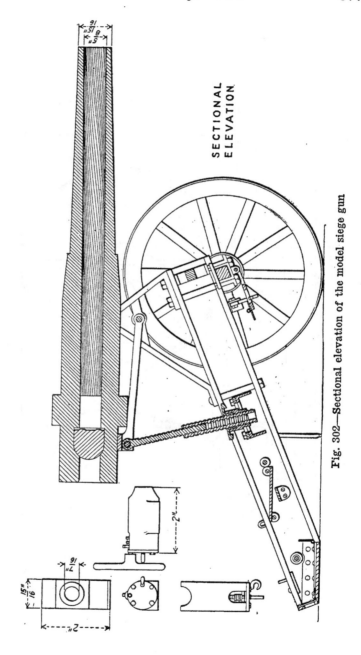

SECTIONAL ELEVATION.

Fig. 302—Sectional elevation of the model siege gun

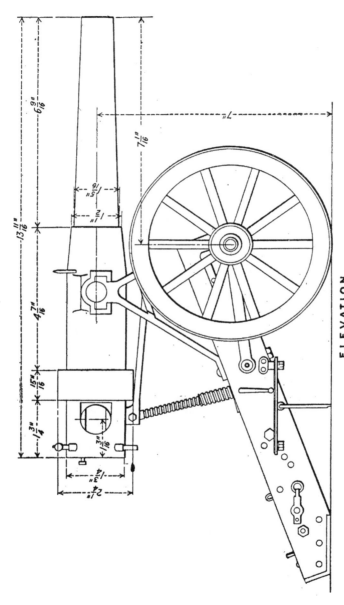

ELEVATION

Fig. 303—General plan and elevation of the model siege gun.

a difficult part of the gun to produce and there would be no objection to making an ordinary bearing to take its place.

The shaft for the wheels is now put in place and the bearings for these are bolted to the upper end of the carriage frame. The wheels are made out of wood and this can easily be done by cutting two semicircular rim pieces with a coping saw. The spokes are cut from oak, and one end is made to fit in the hub, and holes are drilled in the two rim pieces to receive the other end of the spokes. Before the spokes are put in place, the holes should be filled with a good grade of carpenters' glue. After this is done the periphery of the wheel is wound with wire and is left in this way until the glue is firmly set. A small brass bearing is then inserted in the hub and a thin strip of sheet iron is placed around the periphery of the wheel to act as a rim. The rim can be made of the proper diameter, after which it is brought to a red heat and burned on to the wooden wheel. Two small bushings placed on the wheel shaft prevent the wheels from sliding out of their place. Attached to the back end of the carriage is a small scale which drops into position when the gun is being fired. This is graduated in millimeters. Directly under the wheel shaft there is another scale graduated in the same manner, which also drops into position when the gun is fired. These scales are used to bring the gun back to its proper position after it flinches upon being fired. The forward scale is held in place with a hook and this hook releases it and it drops in a vertical position.

This completes the gun, and after all parts are polished it can be painted a dull black, which gives the finished model a very business-like appearance. The paint should have no gloss in it, as this would detract from the appearance of the model.

INDEX

A

CATALOGUE

Of the LATEST and BEST

PRACTICAL and MECHANICAL

BOOKS

Including Automobile and Aviation Books

Any of these books will be sent prepaid to any part of the world, on receipt of price. Remit by Draft, Postal Order, Express Order or Registered Letter

Published and For Sale By

The Norman W. Henley Publishing Co.,

2 West 45th Street, New York, U.S.A.

INDEX

Any of these books promptly sent prepaid to any address in the world on receipt of price.

HOW TO REMIT—By Postal Money Order, Express Money Order, Bank Draft or Registered Letter.

AUTOMOBILES AND MOTORCYCLES

The Modern Gasoline Automobile—Its Design, Construction, and Operation, 1918 Edition. By VICTOR W. PAGÉ, M.S.A.E.

This is the most complete, practical and up-to-date treatise on gasoline automobiles and their component parts ever published. In the new *revised* and *enlarged* 1918 *edition*, all phases of automobile construction, operation and maintenance are fully and completely described, and in language anyone can understand Every part of all types of automobiles, from light cycle-cars to heavy motor trucks and tractors, are described in a thorough manner, not only the automobile, but every item of it; equipment, accessories, tools needed, supplies and spare parts necessary for its upkeep, are fully discussed.

It is clearly and concisely written by an expert familiar with every branch of the automobile industry and the originator of the practical system of self-education on technical subjects. It is a liberal education in the automobile art, useful to all who motor for either business or pleasure.

Anyone reading the incomparable treatise is in touch with all improvements that have been made in motor-car construction. All latest developments, such as high speed aluminum motors and multiple valve and sleeve-valve engines, are considered in detail. The latest ignition, carburetor and lubrication practice is outlined. New forms of change speed gears, and final power transmission systems, and all latest chassis improvements are shown and described. This book is used in all leading automobile schools and is conceded to be the STANDARD TREATISE. The chapter on Starting and Lighting Systems has been greatly enlarged, and many automobile engineering features that have long puzzled laymen are explained so clearly that the underlying principles can be understood by anyone. This book was first published six years ago and so much new matter has been added that it is nearly twice its original size. The only treatise covering various forms of war automobiles and recent developments in motor-truck design as well as pleasure cars. *This book is not too technical for the layman nor too elementary for the more expert. It is an incomparable work of reference for home or school.* 1,000 6x9 pages, nearly 1,000 illustrations, 12 folding plates. Cloth bound. Price **$3.00**

WHAT IS SAID OF THIS BOOK:

"It is the best book on the Automobile seen up to date."—J. H. Pile, Associate Editor *Automobile Trade Journal*.

"Every Automobile Owner has use for a book of this character."—*The Tradesman*.

"This book is superior to any treatise heretofore published on the subject."—*The Inventive Age*.

"We know of no other volume that is so complete in all its departments, and in which the wide field of automobile construction with its mechanical intricacies is so plainly handled, both in the text and in the matter of illustrations."—*The Motorist*.

"The book is very thorough, a careful examination failing to disclose any point in connection with the automobile, its care and repair, to have been overlooked."—*Iron Age*.

"Mr. Pagé has done a great work, and benefit to the Automobile Field."—W. C. Hasford, Mgr. Y. M. C. A. Automobile School, Boston, Mass.

"It is just the kind of a book a motorist needs if he wants to understand his car."—*American Thresherman*.

The Model T Ford Car, Its Construction, Operation and Repair. By VICTOR W. PAGÉ, M.S.A.E.

This is a complete instruction book. All parts of the Ford Model T Car are described and illustrated; the construction is fully described and operating principles made clear to everyone. Every Ford owner needs this practical book. You don't have to guess about the construction or where the trouble is, as it shows how to take all parts apart and how to locate and fix all faults. The writer, Mr. Pagé, has operated a Ford car for many years and writes from actual knowledge. Among the contents are: 1. The Ford Car: Its Parts and Their Functions. 2. The Engine and Auxiliary Groups. How the Engine Works—The Fuel Supply System—The Carburetor—Making the Ignition Spark—Cooling and Lubrication 3. Details of Chassis. Change Speed Gear—Power Transmission—Differential Gear Action—Steering Gear—Front Axle—Frame and Springs—Brakes 4. How to Drive and Care for the Ford. The Control System Explained—Starting the Motor—Driving the Car—Locating Roadside Troubles—Tire Repairs—Oiling the Chassis—Winter Care of Car. 5. Systematic Location of Troubles and Remedies Faults in Engine—Faults in Carburetor—Ignition Troubles—Cooling and Lubrication System Defects—Adjustment of Transmission Gear—General Chassis Repairs. 95 illustrations, 300 pages, 2 large folding plates. Price **$1.00**

How to Run an Automobile. By VICTOR W. PAGÉ, M.S.A.E.

This treatise gives concise instructions for starting and running all makes of gasoline automobiles, how to care for them, and gives distinctive features of control. Describes every step for shifting gears, controlling engines, etc. Among the chapters contained are: I.—Automobile Parts and Their Functions. II.—General Starting and Driving Instructions. III —Typical 1917 Control Systems. IV.—Care of Automobiles. 178 pages. 72 specially made illustrations. Price **$1.00**

Automobile Repairing Made Easy. By Victor W. Pagé, M.S.A.E.

A comprehensive, practical exposition of every phase of modern automobile repairing prac_ tice. Outlines every process incidental to motor car restoration. Gives plans for workshop construction, suggestions for equipment, power needed, machinery and tools necessary to carry on business successfully. Tells how to overhaul and repair all parts of all auto_ mobiles. Everything is explained so simply that motorists and students can acquire a full working knowledge of automobile repairing. This work starts with the engine, then considers carburetion, ignition, cooling and lubrication systems. The clutch, change speed gearing and transmission system are considered in detail. Contains instructions for repairing all types of axles, steering gears and other chassis parts. Many tables, short cuts in figuring and rules of practice are given for the mechanic. Explains fully valve and magneto timing, "tuning" engines, systematic location of trouble, repair of ball and roller bearings, shop kinks, first aid to injured and a multitude of subjects of interest to all in the garage and repair business. *This book contains special instructions on electric starting, lighting and ignition systems, tire repairing and rebuilding, autogenous welding, brazing, heat treatment of steel, latest timing practice, eight and twelve-cylinder motors, etc.* 5¾x8. Cloth. 1,056 pages, 1,000 illustrations, 11 folding plates. Price **$3.00**

WHAT IS SAID OF THIS BOOK:

"'Automobile Repairing Made Easy' is the best book on the subject I have ever seen and the only book I ever saw that is of any value in a garage."—Fred Jeffrey, Martinsburg, Neb. "I wish to thank you for sending me a copy of 'Automobile Repairing Made Easy.' I do not think it could be excelled."—S. W. Gisriel, Director of Instruction, Y. M. C. A., Philadelphia, Pa.

Questions and Answers Relating to Modern Automobile Construction, Driving and Repair. By Victor W. Pagé, M.S.A.E.

A practical self-instructor for students, mechanics and motorists, consisting of thirty-seven lessons in the form of questions and answers, written with special reference to the requirements of the non-technical reader desiring easily understood, explanatory matter relating to all branches of automobiling. The subject-matter is absolutely correct and explained in simple language. If you can't answer all of the following questions, you need this work. The answers to these and over 2,000 more are to be found in its pages Give the name of all important parts of an automobile and describe their functions Describe action of latest types of kerosene carburetors What is the difference between a "double" ignition system and a "dual" ignition system? Name parts of an induction coil How are valves timed? What is an electric motor starter and how does it work? What are advantages of worm drive gearing? Name all important types of ball and roller bearings What is a "three-quarter" floating axle? What is a two-speed axle? What is the Vulcan electric gear shift? Name the causes of lost power in automobiles. Describe all noises due to deranged mechanism and give causes, How can you adjust a carburetor by the color of the exhaust gases? What causes "popping" in the carburetor? What tools and supplies are needed to equip a car? How do you drive various makes of cars? What is a differential lock and where is it used? Name different systems of wire wheel construction, etc, etc. A popular work at a popular price. 5¾x7½. Cloth. 650 pages, 350 illustrations, 3 folding plates Price **$1.50**

WHAT IS SAID OF THIS BOOK:

"If you own a car—get this book."—*The Glassworker*.
"Mr Page has the faculty of making difficult subjects plain and understandable."—*Bristol Press*.
"We can name no writer better qualified to prepare a book of instruction on automobiles than Mr. Victor W. Pagé."—*Scientific American*.
"The best automobile catechism that has appeared"—*Automobile Topics*.
"There are few men, even with long experience, who will not find this book useful Great pains have been taken to make it accurate. Special recommendation must be given to the illustrations, which have been made specially for the work. Such excellent books as this greatly assist in fully understanding your automobile."—*Engineering News*.

The Automobilist's Pocket Companion and Expense Record. Arranged by Victor W. Pagé, M.S.A.E.

This book is not only valuable as a convenient cost record but contains much information of value to motorists Includes a condensed digest of auto laws of all States, a lubrication schedule, hints for care of storage battery and car, of tires, location of road troubles, anti-freezing solutions, horse-power table, driving hints and many useful tables and recipes of interest to all motorists Not a technical book in any sense of the word, just a collection of practical facts in simple language for the everyday motorist. Price **$1.00**

Modern Starting, Lighting and Ignition Systems. By Victor W. Pagé, M.E.

This practical volume has been written with special reference to the requirements of the non-technical reader desiring easily understood, explanatory matter, relating to all types of automobile ignition, starting and lighting systems. It can be understood by anyone, even without electrical knowledge, because elementary electrical principles are considered before any attempt is made to discuss features of the various systems. These basic principles are clearly stated and illustrated with simple diagrams. *All the leading systems of starting, lighting and ignition have been described and illustrated with the co-operation of the experts employed by the manufacturers.* Wiring diagrams are shown in both technical and non-technical forms All symbols are fully explained. It is a comprehensive review of modern starting and ignition system practice, and includes a complete exposition of storage battery construction, care and repair. All types of starting motors, generators, magnetos, and all ignition or lighting system-units are fully explained. *Every person in the automobile business needs this volume.* Among some of the subjects treated are: I.—Elementary Electricity; Current Production; Flow; Circuits; Measurements; Definitions; Magnetism; Battery Action; Generator Action II —Battery Ignition Systems. III.—Magneto Ignition Systems. IV.—Elementary Exposition of Starting System Principles V —Typical Starting and Lighting Systems; Practical Application; Wiring Diagrams; Auto-lite, Bijur, Delco, Dyneto-Entz, Gray and Davis, Remy, U. S. L , Westinghouse, Bosch-Rushmore, Genemotor, North-East, etc. VI —Locating and Repairing Troubles in Starting and Lighting Systems. VII.—Auxiliary. Electric Systems; Gear-shifting by Electricity; Warning Signals; Electric Brake; Entz-Transmission, Wagner-Saxon Circuits, Wagner-Studebaker Circuits. 5¼x7½. Cloth. 530 pages, 297 illustrations, 3 folding plates.
Price **$1.50**

Automobile Welding With the Oxy-Acetylene Flame. By M. Keith Dunham.

This is the only complete book on the "why" and "how" of Welding with the Oxy-Acetylene Flame, and from its pages one can gain information so that he can weld anything that comes along
No one can afford to be without this concise book, as it first explains the apparatus to be used, and then covers in detail the actual welding of all automobile parts. The welding of aluminum, cast iron, steel, copper, brass and malleable iron is clearly explained, as well as the proper way to burn the carbon out of the combustion head of the motor. Among the contents are: Chapter I.—Apparatus Knowledge. Chapter II.—Shop Equipment and Initial Procedure. Chapter III.—Cast Iron. Chapter IV.—Aluminum. Chapter V.—Steel. Chapter VI.—Malleable Iron, Copper, Brass, Bronze. Chapter VII.—Carbon Burning and other Uses of Oxygen and Acetylene. Chapter VIII.—How to Figure Cost of Welding. 167 pages, fully illustrated. Price **$1.00**

Storage Batteries Simplified. By Victor W. Pagé, M.S.A.E.

A comprehensive treatise devoted entirely to secondary batteries and their maintenance, repair and use
This is the most up-to-date book on this subject. Describes fully the Exide, Edison, Gould, Willard, U. S L and other storage battery forms in the types best suited for automobile, stationary and marine work. Nothing of importance has been omitted that the reader should know about the practical operation and care of storage batteries. No details have been slighted The instructions for charging and care have been made as simple as possible. Brief Synopsis of Chapters: Chapter I —Storage Battery Development; Types of Storage Batteries; Lead Plate Types; The Edison Cell. Chapter II.—Storage Battery Construction; Plates and Girds; Planté Plates; Fauré Plates; Non-Lead Plates; Commercial Battery Designs. Chapter III.—Charging Methods; Rectifiers; Converters; Rheostats; Rules for Charging Chapter IV.—Battery Repairs and Maintenance. Chapter V.—Industrial Application of Storage Batteries; Glossary of Storage Battery Terms. 208 Pages. Very Fully Illustrated. Price . . . . **$1.50 net.**

Motorcycles, Side Cars and Cyclecars; their Construction, Management and Repair. By Victor W. Pagé, M.S.A.E.

The only complete work published for the motorcyclist and cyclecarist Describes fully all leading types of machines, their design, construction, maintenance, operation and repair This treatise outlines fully the operation of two- and four-cycle power plants and all ignition, carburetion and lubrication systems in detail. Describes all representative types of free engine clutches, variable speed gears and power transmission systems Gives complete instructions for operating and repairing all types Considers fully electric self-starting and lighting systems, all types of spring frames and spring forks and shows leading control methods. For those desiring technical information a complete series of tables and many formulæ to assist in designing are included The work tells how to figure power needed to climb grades, overcome air resistance and attain high speeds It shows how to select gear ratios for various weights and powers, how to figure braking efficiency required, gives sizes of belts and chains to transmit power safely, and shows how to design sprockets, belt pulleys, etc This work also includes complete formulæ for figuring horse-power, shows how dynamometer tests are made, defines relative efficiency of air and water-cooled engines, plain and anti-friction bearings and many other data of a practical, helpful, engineering nature. Remember that you get this information in addition to the practical description and instructions which alone are worth several times the price of the book. 350 pages. 350 specially made illustrations, 5 folding plates. Cloth. Price **$1.50**

WHAT IS SAID OF THIS BOOK:

"Here is a book that should be in the cycle repairer's kit."—*American Blacksmith.*
"The best way for any rider to thoroughly understand his machine, is to get a copy of this book; it is worth many times its price."—*Pacific Motorcyclist.*

AUTOMOBILE AND MOTORCYCLE CHARTS

Chart. Location of Gasoline Engine Troubles Made Easy—A Chart Showing Sectional View of Gasoline Engine. Compiled by VICTOR W. PAGÉ, M.S.A.E.

It shows clearly all parts of a typical four-cylinder gasoline engine of the four-cycle type. It outlines distinctly all parts liable to give trouble and also details the derangements apt to interfere with smooth engine operation.
Valuable to students, motorists, mechanics, repairmen, garagemen, automobile salesmen, chauffeurs, motorboat owners, motor-truck and tractor drivers, aviators, motor-cyclists, and all others who have to do with gasoline power plants.
It simplifies location of all engine troubles, and while it will prove invaluable to the novice, it can be used to advantage by the more expert. It should be on the walls of every public and private garage, automobile repair shop, club house or school. It can be carried in the automobile or pocket with ease, and will insure against loss of time when engine trouble manifests itself.
This sectional view of engine is a complete review of all motor troubles. It is prepared by a practical motorist for all who motor. More information for the money than ever before offered. No details omitted. Size 25x38 inches. Securely mailed on receipt of   **25 cents**

Chart. Location of Ford Engine Troubles Made Easy. Compiled by VICTOR W. PAGÉ, M.S.A.E.

This shows clear sectional views depicting all portions of the Ford power plant and auxiliary groups. It outlines clearly all parts of the engine, fuel supply system, ignition group and cooling system, that are apt to give trouble, detailing all derangements that are liable to make an engine lose power, start hard or work irregularly. This chart is valuable to students, owners, and drivers, as it simplifies location of all engine faults. Of great advantage as an instructor for the novice, it can be used equally well by the more expert as a work of reference and review. It can be carried in the tool-box or pocket with ease and will save its cost in labor eliminated the first time engine trouble manifests itself. Prepared with special reference to the average man's needs and is a practical review of all motor troubles because it is based on the actual experience of an automobile engineer-mechanic with the mechanism the chart describes. It enables the non-technical owner or operator of a Ford car to locate engine derangements by systematic search, guided by easily recognized symptoms instead of by guesswork. It makes the average owner independent of the roadside repair shop when touring. Must be seen to be appreciated. Size 25x38 inches. Printed on heavy bond paper. Price . . . .   **25 cents**

Chart. Lubrication of the Motor Car Chassis. Compiled by VICTOR W. PAGÉ, M.S.A.E.

This chart presents the plan view of a typical six-cylinder chassis of standard design and all parts are clearly indicated that demand oil, also the frequency with which they must be lubricated and the kind of oil to use. A practical chart for all interested in motor-car maintenance. Size 24x38 inches. Price . . . .   **25 cents**

Chart. Location of Carbureton Troubles Made Easy. Compiled by VICTOR W. PAGÉ, M.S.A.E.

This chart shows all parts of a typical pressure feed fuel supply system and gives causes of trouble, how to locate defects and means of remedying them. Size 24x38 inches. Price . . . .   **25 cents**

Chart. Location of Ignition System Troubles Made Easy. Compiled by VICTOR W. PAGÉ, M.S.A.E.

In this diagram all parts of a typical double ignition system using battery and magneto current are shown, and suggestions are given for readily finding ignition troubles and eliminating them when found. Size 24x38 inches. Price . . . .   **25 cents**

Chart. Location of Cooling and Lubrication System Faults. Compiled by VICTOR W. PAGÉ, M.S.A.E.

This composite diagram shows a typical automobile power p ant using pump circulated water-cooling system and the most popular lubrication method.l Gives suggestions for curing all overheating and loss of power faults due to faulty action of the oiling or cooling group. Size 24x38 inches. Price . . . . **25 cents**

Chart. Motorcycle Troubles Made Easy. Compiled by VICTOR W. PAGÉ, M.S.A.E.

A chart showing sectional view of a single-cylinder gasoline engine. This chart simplifies location of all power-plant troubles. A single-cylinder motor is shown for simplicity It outlines distinctly all parts liable to give trouble and also details the derangements apt to interfere with smooth engine operation. This chart will prove of value to all who have to do with the operation, repair or sale of motorcycles. No details omitted. Size 30x20 inches. Price . . . . **25 cents**

AVIATION

Aviation Engines, their Design, Construction, Operation and Repair. By Lieut. VICTOR W. PAGÉ, Aviation Section, S.C.U.S.R.

A practical work containing valuable instructions for aviation students, mechanicians, squadron engineering officers and all interested in the construction and upkeep of airplane power plants.

The rapidly increasing interest in the study of aviation, and especially of the highly developed internal combustion engines that make mechanical flight possible, has created a demand for a text-book suitable for schools and home study that will clearly and concisely explain the workings of the various aircraft engines of foreign and domestic manufacture.

This treatise, written by a recognized authority on all of the practical aspects of internal combustion engine construction, maintenance and repair fills the need as no other book does. The matter is logically arranged; all descriptive matter is simply expressed and copiously illustrated so that anyone can understand airplane engine operation and repair even if without previous mechanical training. This work is invaluable for anyone desiring to become an aviator or aviation mechanician.

The latest rotary types, such as the Gnome, Monosoupape, and Le Rhone, are fully explained, as well as the recently developed Vee and radial types. The subjects of carburetion, ignition, cooling and lubrication also are covered in a thorough manner. The chapters on repair and maintenance are distinctive and found in no other book on this subject.

Invaluable to the student, mechanic and soldier wishing to enter the aviation service. Not a technical book, but a practical, easily understood work of reference for all interested in aeronautical science. 576 octavo pages. 253 specially made engravings. Price . **$3.00 net**

GLOSSARY OF AVIATION TERMS

Termes D'Aviation, English-French, French-English. Compiled by Lieuts. VICTOR W. PAGÉ, A.S., S.C.U.S.R., and PAUL MONTARIOL of the French Flying Corps, on duty on Signal Corps Aviation School, Mineola, L. I.

A complete, well illustrated volume intended to facilitate conversation between English-speaking and French aviators. A very valuable book for all who are about to leave for duty overseas.

Approved for publication by Major W. G. Kilner, S.C., U.S.C.O. Signal Corps Aviation School. Hazlehurst Field, Mineola, L. I.

This book should be in every Aviator's and Mechanic's Kit for ready reference. 128 pages. Fully illustrated with detailed engravings. Price **$1.00**

Aviation Chart. Location of Airplane Power Plant Troubles Made Easy. By Lieut. VICTOR W. PAGÉ, A.S., S.C.U.S.R.

A large chart outlining all parts of a typical airplane power plant, showing the points where trouble is apt to occur and suggesting remedies for the common defects. Intended especially for Aviators and Aviation Mechanics on School and Field Duty. Price . . **50 cents**

BRAZING AND SOLDERING

Brazing and Soldering. By JAMES F. HOBART.

The only book that shows you just how to handle any job of brazing or soldering that comes along; it tells you what mixture to use, how to make a furnace if you need one. Full of valuable kinks. The fifth edition of this book has just been published, and to it much new matter and a large number of tested formulæ for all kinds of solders and fluxes have been added. Illustrated. Price . ., ... **25 cents**

CHARTS

Aviation Chart. Location of Airplane Power Plant Troubles Made Easy.
By Lieut. VICTOR W. PAGÉ, A.S., S.C.U.S.R.

A large chart outlining all parts of a typical airplane power plant, showing the points where trouble is apt to occur and suggesting remedies for the common defects. Intended especially for Aviators and Aviation Mechanics on School and Field Duty. Price . . . . **50 cents**

Gasoline Engine Troubles Made Easy—A Chart Showing Sectional View of Gasoline Engine. Compiled by Lieut. VICTOR W. PAGÉ, A.S., S.C.U.S.R.

It shows clearly all parts of a typical four-cylinder gasoline engine of the four-cycle type. It outlines distinctly all parts liable to give trouble and also details the derangements apt to interfere with smooth engine operation.

Valuable to students, motorists, mechanics, repairmen, garagemen, automobile salesmen, chauffeurs, motor-boat owners, motor-truck and tractor drivers, aviators, motor-cyclists, and all others who have to do with gasoline power plants.

It simplifies location of all engine troubles, and while it will prove invaluable to the novice, it can be used to advantage by the more expert. It should be on the walls of every public and private garage, automobile repair shop, club house or school. It can be carried in the automobile or pocket with ease and will insure against loss of time when engine trouble manifests itself.

This sectional view of engine is a complete review of all motor troubles. It is prepared by a practical motorist for all who motor. No details omitted. Size 25x38 inches. Price   **25 cents**

Lubrication of the Motor Car Chassis.

This chart presents the plan view of a typical six-cylinder chassis of standard design and all parts are clearly indicated that demand oil, also the frequency with which they must be lubricated and the kind of oil to use. A practical chart for all interested in motor-car maintenance. Size 24x38 inches. Price . . . . **25 cents**

Location of Carburetion Troubles Made Easy.

This chart shows all parts of a typical pressure feed fuel supply system and gives causes of trouble, how to locate defects and means of remedying them. Size 24x38 inches. Price . . . . **25 cents**

Location of Ignition System Troubles Made Easy.

In this chart all parts of a typical double ignition system using battery and magneto current are shown and suggestions are given for readily finding ignition troubles and eliminating them when found. Size 24x38 inches. Price . . . . **25 cents**

Location of Cooling and Lubrication System Faults.

This composite chart shows a typical automobile power plant using pump circulated water-cooling system and the most popular lubrication method. Gives suggestions for curing all overheating and loss of power faults due to faulty action of the oiling or cooling group. Size 24x38 inches. Price . . . . **25 cents**

Motorcycle Troubles Made Easy—A Chart Showing Sectional View of Single-Cylinder Gasoline Engine. Compiled by VICTOR W. PAGÉ, M.S.A.E.

This chart simplifies location of all power-plant troubles, and will prove invaluable to all who have to do with the operation, repair or sale of motorcycles. No details omitted. Size 25x38 inches. Price . . . . **25 cents**

Location of Ford Engine Troubles Made Easy.   Compiled by VICTOR W. PAGÉ, M.S.A.E.

This shows clear sectional views depicting all portions of the Ford power plant and auxiliary groups. It outlines clearly all parts of the engine, fuel supply system, ignition group and cooling system, that are apt to give trouble, detailing all derangements that are liable to make an engine lose power, start hard or work irregularly. This chart is valuable to students, owners, and drivers, as it simplifies location of all engine faults. Of great advantage as an instructor for the novice, it can be used equally well by the more expert as a work of reference and review. It can be carried in the toolbox or pocket with ease and will save its cost in labor eliminated the first time engine trouble manifests itself. Prepared with special reference to the average man's needs and is a practical review of all motor troubles because it is based on the actual experience of an automobile engineer-mechanic with the mechanism the chart describes. It enables the non-technical owner or operator of a Ford car to locate engine derangements by systematic search, guided by easily recognized symptoms instead of by guesswork. It makes the average owner independent of the roadside repair shop when touring. Must be seen to be appreciated. Size 25x38 inches. Printed on heavy bond paper. Price . . . . **25 cents**

Modern Submarine Chart—with Two Hundred Parts Numbered and Named.

A cross-section view, showing clearly and distinctly all the interior of a Submarine of the latest type You get more information from this chart, about the construction and operation of a Submarine, than in any other way. No details omitted—everything is accurate and to scale. It is absolutely correct in every detail, having been approved by Naval Engineers. All the machinery and devices fitted in a modern Submarine Boat are shown, and to make the engraving more readily understood all the features are shown in operative form, with Officers and Men in the act of performing the duties assigned to them in service conditions. This CHART IS REALLY AN ENCYCLOPEDIA OF A SUBMARINE. It is educational and worth many times its cost. Mailed in a Tube for . . . . . **25 cents**

Box Car Chart.

A chart showing the anatomy of a box car, having every part of the car numbered and its proper name given in a reference list. Price . . . . **25 cents**

Gondola Car Chart.

A chart showing the anatomy of a gondola car, having every part of the car numbered and its proper reference name given in a reference list. Price . . . . **25 cents**

Passenger-Car Chart.

A chart showing the anatomy of a passenger-car, having every part of the car numbered and its proper name given in a reference list . . . . **25 cents**

Steel Hopper Bottom Coal Car.

A chart showing the anatomy of a steel Hopper Bottom Coal Car, having every part of the car numbered and its proper name given in a reference list. Price . . . . . . **25 cents**

Tractive Power Chart.

A chart whereby you can find the tractive power or drawbar pull of any locomotive without making a figure. Shows what cylinders are equal, how driving wheels and steam pressure affect the power. What sized engine you need to exert a given drawbar pull or anything you desire in this line. Price . . . . **50 cents**

Horse-Power Chart.

Shows the horse-power of any stationary engine without calculation No matter what the cylinder diameter of stroke, the steam pressure of cut-off, the revolutions, or whether condensing or non-condensing, it's all there. Easy to use, accurate, and saves time and calculations. Especially useful to engineers and designers. Price . . . . **50 cents**

Boiler Room Chart. By Geo. L. Fowler.

A chart—size 14x28 inches—showing in isometric perspective the mechanisms belonging in a modern boiler room The various parts are shown broken or removed, so that the internal construction is fully illustrated Each part is given a reference number, and these, with the corresponding name, are given in a glossary printed at the sides This chart is really a dictionary of the boiler room—the names of more than 200 parts being given. Price . **25 cents**

COKE

Modern Coking Practice, Including Analysis of Materials and Products.
By J. E. Christopher and T. H. Byrom.

This, the standard work on the subject, has just been revised. It is a practical work for those engaged in Coke manufacture and the recovery of By-products Fully illustrated with folding plates It has been the aim of the authors, in preparing this book, to produce one which shall be of use and benefit to those who are associated with, or interested in, the modern developments of the industry. Among the Chapters contained in Volume I are· Introduction; Classification of Fuels; Impurities of Coals, Coal Washing; Sampling and Valuation of Coals, etc ; Power of Fuels; History of Coke Manufacture; Developments in the Coke Oven Design; Recent Types of Coke Ovens; Mechanical Appliances at Coke Ovens; Chemical and Physical Examination of Coke. Volume II covers fully the subject of By-Products. Price, per volume . . . . **$3.00 net**

COMPRESSED AIR

Compressed Air in All Its Applications. By GARDNER D. HISCOX.

This is the most complete book on the subject of Air that has ever been issued, and its thirty-five chapters include about every phase of the subject one can think of. It may be called an encyclopedia of compressed air. It is written by an expert, who, in its 665 pages, has dealt with the subject in a comprehensive manner, no phase of it being omitted. Includes the physical properties of air from a vacuum to its highest pressure, its thermodynamics, compression, transmission and uses as a motive power, in the Operation of Stationary and Portable Machinery, in Mining, Air Tools, Air Lifts, Pumping of Water, Acids, and Oils; the Air Blast for Cleaning and Painting the Sand Blast and its Work, and the Numerous Appliances in which Compressed Air is a Most Convenient and Economical Transmitter of Power for Mechanical Work, Railway Propulsion, Refrigeration, and the Various Uses to which Compressed Air has been applied. Includes forty-four tables of the physical properties of air, its compression, expansion, and volumes required for various kinds of work, and a list of patents on compressed air from 1875 to date. Over 500 illustrations, 5th Edition, revised and enlarged

Cloth bound. Price. **$5.00**
Half Morocco. Price **$6.50**

CONCRETE

Concrete Workers' Reference Books. A Series of Popular Handbooks for Concrete Users. Prepared by A. A. HOUGHTON. . . . . .50 cents

The author, in preparing this Series, has not only treated on the usual types of construction, but explains and illustrates molds and systems that are not patented, but which are equal in value and often superior to those restricted by patents. These molds are very easily and cheaply constructed and embody simplicity, rapidity of operation, and the most successful results in the molded concrete. Each of these books is fully illustrated, and the subjects are exhaustively treated in plain English.

Concrete Wall Forms. By A. A. HOUGHTON.

A new automatic wall clamp is illustrated with working drawings. Other types of wall forms, clamps, separators, etc., are also illustrated and explained. (No. 1 of Series) Price **50 cents**

Concrete Floors and Sidewalks. By A. A. HOUGHTON.

The molds for molding squares, hexagonal and many other styles of mosaic floor and sidewalk blocks are fully illustrated and explained. (No. 2 of Series) Price . . . . . .50 cents

Practical Concrete Silo Construction. By A. A. HOUGHTON.

Complete working drawings and specifications are given for several styles of concrete silos, with illustrations of molds for monolithic and block silos. The tables, data, and information presented in this book are of the utmost value in planning and constructing all forms of concrete silos. (No. 3 of Series) Price . . . . **50 cents**

Molding Concrete Chimneys, Slate and Roof Tiles. By A. A. HOUGHTON.

The manufacture of all types of concrete slate and roof tile is fully treated Valuable data on all forms of reinforced concrete roofs are contained within its pages. The construction of concrete chimneys by block and monolithic systems is fully illustrated and described. A number of ornamental designs of chimney construction with molds are shown in this valuable treatise. (No. 4 of Series.) Price . . . . **50 cents**

Molding and Curing Ornamental Concrete. By A. A. HOUGHTON.

The proper proportions of cement and aggregates for various finishes, also the method of thoroughly mixing and placing in the molds, are fully treated An exhaustive treatise on this subject that every concrete worker will find of daily use and value. (No. 5 of Series.) Price . . . . **50 cents**

Concrete Monuments, Mausoleums and Burial Vaults. By A. A. HOUGHTON.

The molding of concrete monuments to imitate the most expensive cut stone is explained in this treatise, with working drawings of easily built molds. Cutting inscriptions and designs are also fully treated. (No. 6 of Series.) Price . . . . **50 cents**

Molding Concrete Bathtubs, Aquariums and Natatoriums. By A. A. HOUGHTON.

Simple molds and instruction are given for molding many styles of concrete bathtubs, swimming-pools, etc. These molds are easily built and permit rapid and successful work. (No. 7 of Series.) Price . . . . **50 cents**

Concrete Bridges, Culverts and Sewers. By A. A. HOUGHTON.

A number of ornamental concrete bridges with illustrations of molds are given. A collapsible center or core for bridges, culverts and sewers is fully illustrated with detailed instructions for building. (No. 8 of Series.) Price . . . . **50 cents**

Constructing Concrete Porches. By A. A. HOUGHTON.

A number of designs with working drawings of molds are fully explained so any one can easily construct different styles of ornamental concrete porches without the purchase of expensive molds. (No. 9 of Series.) Price . . . . **50 cents**

Molding Concrete Flower-Pots, Boxes, Jardinieres, Etc. By A. A. HOUGHTON.

The molds for producing many original designs of flower-pots, urns, flower-boxes, jardinieres, etc., are fully illustrated and explained, so the worker can easily construct and operate same. (No. 10 of Series.) Price . . . . **50 cents**

Molding Concrete Fountains and Lawn Ornaments. By A. A. HOUGHTON.

The molding of a number of designs of lawn seats, curbing, hitching posts, pergolas, sun dials and other forms of ornamental concrete for the ornamentation of lawns and gardens, is fully illustrated and described. (No. 11 of Series.) Price . . . . **50 cents**

Concrete from Sand Molds. By A. A. HOUGHTON.

A Practical Work treating on a process which has heretofore been held as a trade secret by the few who possessed it, and which will successfully mold every and any class of ornamental concrete work. The process of molding concrete with sand molds is of the utmost practical value, possessing the manifold advantages of a low cost of molds, the ease and rapidity of operation, perfect details to all ornamental designs, density and increased strength of the concrete, perfect curing of the work without attention and the easy removal of the molds regardless of any undercutting the design may have. 192 pages. Fully illustrated. Price **$2.00**

Ornamental Concrete without Molds. By A. A. HOUGHTON.

The process for making ornamental concrete without molds has long been held as a secret, and now, for the first time, this process is given to the public. The book reveals the secret and is the only book published which explains a simple, practical method whereby the concrete worker is enabled, by employing wood and metal templates of different designs, to mold or model in concrete any Cornice, Archivolt, Column, Pedestal, Base Cap, Urn or Pier in a monolithic form—right upon the job. These may be molded in units or blocks and then built up to suit the specifications demanded. This work is fully illustrated, with detailed engravings. Price **$2.00**

Concrete for the Farm and in the Shop. By H. COLIN CAMPBELL, C.E., E.M.

"Concrete for the Farm and in the Shop" is a new book from cover to cover, illustrating and describing in plain, simple language many of the numerous applications of concrete within the range of the home worker. Among the subjects treated are: Principles of Reinforcing; Methods of Protecting Concrete so as to Insure Proper Hardening; Home-made Mixers; Mixing by Hand and Machine; Form Construction, Described and Illustrated by Drawings and Photographs; Construction of Concrete Walls and Fences; Concrete Fence Posts; Concrete Gate Posts; Corner Posts; Clothes Line Posts; Grape Arbor Posts; Tanks; Troughs; Cisterns; Hog Wallows; Feeding Floors and Barnyard Pavements; Foundations; Well Curbs and Platforms; Indoor Floors; Sidewalks; Steps; Concrete Hotbeds and Cold Frames; Concrete Slab Roofs; Walls for Buildings; Repairing Leaks in Tanks and Cisterns; and all topics associated with these subjects as bearing upon securing the best results from concrete are dwelt upon at sufficient length in plain every-day English so that the inexperienced person desiring to undertake a piece of concrete construction can, by following the directions set forth in this book, secure 100 per cent. success every time. A number of convenient and practical tables for estimating quantities, and some practical examples, are also given. (5x7.) 149 pages. 51 illustrations. Price . . . . **75 cents**

Popular Handbook for Cement and Concrete Users. By MYRON H. LEWIS.

This is a concise treatise of the principles and methods employed in the manufacture and use of cement in all classes of modern works. The author has brought together in this work all the salient matter of interest to the user of concrete and its many diversified products. The matter is presented in logical and systematic order, clearly written, fully illustrated and free from involved mathematics. Everything of value to the concrete user is given, including kinds of cement employed in construction, concrete architecture, inspection and testing, waterproofing, coloring and painting, rules, tables, working and cost data. The book comprises thirty-three chapters, as follow: Introductory. Kinds of Cement and How They are Made. Properties. Testing and Requirements of Hydraulic Cement. Concrete and Its Properties. Sand, Broken Stone and Gravel for Concrete. How to Proportion the Materials. How to Mix and Place Concrete. Forms of Concrete Construction. The Architectural and Artistic Possibilities of Concrete. Concrete Residences. Mortars, Plasters and Stucco, and How to Use Them. The Artistic Treatment of Concrete Surfaces. Concrete Building

Blocks. The Making of Ornamental Concrete. Concrete Pipes, Fences, Posts, etc. Essen-tial Features and Advantages of Reenforced Concrete. How to Design Reenforced Con-crete Beams, Slabs and Columns. Explanations of the Methods and Principles in Designing Reenforced Concrete, Beams and Slabs. Systems of Reenforcement Employed. Reen-forced Concrete in Factory and General Building Construction. Concrete in Foundation Work. Concrete Retaining Walls, Abutments and Bulkheads. Concrete Arches and Arch Bridges. Concrete Beam and Girder Bridges. Concrete in Sewerage and Draining Works   Concrete Tanks, Dams and Reservoirs   Concrete Sidewalks, Curbs and Pavements. Concrete in Railroad Construction. The Utility of Concrete on the Farm. The Waterproofing of Con-crete Structures. Grout of Liquid Concrete and Its Use. Inspection of Concrete Work. Cost of Concrete Work. Some of the special features of the book are: 1.—The Attention Paid to the Artistic and Architectural Side of Concrete Work. 2.—The Authoritative Treat-ment of the Problem of Waterproofing Concrete. 3.—An Excellent Summary of the Rules to be Followed in Concrete Construction. 4.—The Valuable Cost Data and Useful Tables given. A valuable Addition to the Library of Every Cement and Concrete User. Price . **$2.50**

WHAT IS SAID OF THIS BOOK:

"The field of Concrete Construction is well covered and the matter contained is well within the understanding of any person."—*Engineering-Contracting.*
"Should be on the bookshelves of every contractor, engineer, and architect in the land."—*National Builder.*

Waterproofing Concrete. By MYRON H. LEWIS.

Modern Methods of Waterproofing Concrete and Other Structures. A condensed statement of the Principles, Rules, and Precautions to be Observed in Waterproofing and Dampproofing Structures and Structural Materials. Paper binding. Illustrated. Price . . . . **50 cents**

DICTIONARIES

Aviation Terms, Termes D'Aviation, English-French, French-English.
Compiled by Lieuts. VICTOR W. PAGÉ, A.S., S.C.U.S.R., and PAUL MON-TARIOL, of the French Flying Corps, on duty on Signal Corps Aviation School, Mineola, L. I.

The lists contained are confined to essentials, and special folding plates are included to show all important airplane parts. The lists are divided in four sections as follows: 1.—Flying Field Terms. 2.—The Airplane. 3.—The Engine. 4.—Tools and Shop Terms.
A complete, well illustrated volume intended to facilitate conversation between English-speak-ing and French aviators. A very valuable book for all who are about to leave for duty over-seas.
Approved for publication by Major W. G. Kilner, S.C., U.S C O. Signal Corps Aviation School, Hazelhurst Field, Mineola, L. I.   This book should be in every Aviator's and Mechanic's Kit for ready reference. 128 pages, fully illustrated, with detailed engravings. Price . . **$1.00**

Standard Electrical Dictionary. By T. O'CONOR SLOANE.

An indispensable work to all interested in electrical science. Suitable alike for the student and professional. A practical handbook of reference containing definitions of about 5,000 distinct words, terms and phrases   The definitions are terse and concise and include every term used in electrical science. Recently issued. An entirely new edition   Should be in the possession of all who desire to keep abreast with the progress of this branch of science. Complete, concise and convenient. 682 pages, 393 illustrations. Price **$3.00**

DIES—METAL WORK

Dies: Their Construction and Use for the Modern Working of Sheet Metals.
By J. V. WOODWORTH.

A most useful book, and one which should be in the hands of all engaged in the press working of metals; treating on the Designing, Constructing, and Use of Tools, Fixtures and Devices, together with the manner in which they should be used in the Power Press, for the cheap and rapid production of the great variety of sheet-metal articles now in use   It is designed as a guide to the production of sheet-metal parts at the minimum of cost with the maximum of output   The hardening and tempering of Press tools and the classes of work which may be produced to the best advantage by the use of dies in the power press are fully treated. Its 515 illustrations show dies, press fixtures and sheet-metal working devices, the descriptions of which are so clear and practical that all metal-working mechanics will be able to understand how to design, construct and use them   Many of the dies and press fixtures treated were either constructed by the author or under his supervision. Others were built by skilful mechanics and are in use in large sheet-metal establishments and machine shops. 6th Revised and Enlarged Edition. Price **$3.00**

Punches, Dies and Tools for Manufacturing in Presses. By J. V. WOOD-WORTH.

This work is a companion volume to the author's elementary work entitled "Dies: Their Construction and Use." It does not go into the details of die-making to the extent of the author's previous book, but gives a comprehensive review of the field of operations carried on by presses. A large part of the information given has been drawn from the author's personal experience It might well be termed an Encyclopedia of Die-Making, Punch-Making, Die-Sinking, Sheet-Metal Working, and Making of Special Tools, Sub-presses, Devices and Mechanical Combinations for Punching, Cutting, Bending, Forming, Piercing, Drawing, Compressing and Assembling Sheet-Metal Parts, and also Articles of other Materials in Machine Tools. 2d Edition. Price **$4.00**

Drop Forging, Die-Sinking and Machine-Forming of Steel. By J. V. WOODWORTH.

This is a practical treatise on Modern Shop Practice, Processes, Methods, Machine Tools, and Details treating on the Hot and Cold Machine-Forming of Steel and Iron into Finished Shapes: together with Tools, Dies, and Machinery involved in the manufacture of Duplicate Forgings and Interchangeable Hot and Cold Pressed Parts from Bar and Sheet Metal. This book fills a demand of long standing for information regarding drop-forgings, die-sinking and machine-forming of steel and the shop practice involved, as it actually exists in the modern drop-forging shop. The processes of die-sinking and force-making, which are thoroughly described and illustrated in this admirable work, are rarely to be found explained in such a clear and concise manner as is here set forth. The process of die-sinking relates to the engraving or sinking of the female or lower dies, such as are used for drop-forgings, hot and cold machine-forming, swedging, and the press working of metals. The process of force-making relates to the engraving or raising of the male or upper dies used in producing the lower dies for the press-forming and machine-forging of duplicate parts of metal.

In addition to the arts above mentioned the book contains explicit information regarding the drop-forging and hardening plants, designs, conditions, equipment, drop hammers, forging machines, etc , machine forging, hydraulic forging, autogenous welding and shop practice. The book contains eleven chapters, and the information contained in these chapters is just what will prove most valuable to the forged-metal worker. All operations described in the work are thoroughly illustrated by means of perspective half-tones and outline sketches of the machinery employed. 300 detailed illustrations. Price **$2.50**

DRAWING—SKETCHING PAPER

Practical Perspective. By RICHARDS and COLVIN.

Shows just how to make all kinds of mechanical drawings in the only practical perspective isometric. Makes everything plain, so that any mechanic can understand a sketch or drawing in this way. Saves time in the drawing room, and mistakes in the shops. Contains practical examples of various classes of work. 4th Edition. Price . . . . **50 cents**

Linear Perspective Self-Taught. By HERMAN T. C. KRAUS.

This work gives the theory and practice of linear perspective, as used in architectural, engineering and mechanical drawings. Persons taking up the study of the subject by themselves will be able, by the use of the instruction given, to readily grasp the subject, and by reasonable practice become good perspective draftsmen. The arrangement of the book is good; the plate is on the left-hand, while the descriptive text follows on the opposite page, so as to be readily referred to The drawings are on sufficiently large scale to show the work clearly and are plainly figured. There is included a self-explanatory chart which gives all information necessary for the thorough understanding of perspective This chart alone is worth many times over the price of the book. 2d Revised and Enlarged Edition. Price **$2.50**

Self-Taught Mechanical Drawing and Elementary Machine Design. By F. L. SYLVESTER, M.E., Draftsman, with additions by ERIK OBERG, associate editor of "Machinery."

This is a practical treatise on Mechanical Drawing and Machine Design, comprising the first principles of geometric and mechanical drawing, workshop mathematics, mechanics, strength of materials and the calculations and design of machine details. The author's aim has been to adapt this treatise to the requirements of the practical mechanic and young draftsman and to present the matter in as clear and concise a manner as possible. To meet the demands of this class of students, practically all the important elements of machine design have been dealt with, and in addition algebraic formulas have been explained, and the elements of trigonometry treated in the manner best suited to the needs of the practical man The book is divided into 20 chapters, and in arranging the material, mechanical drawing, pure and simple, has been taken up first, as a thorough understanding of the principles of representing objects facilitates the further study of mechanical subjects. This is followed by the mathematics necessary for the solution of the problems in machine design which are presented later, and a practical introduction to theoretical mechanics and the strength of materials The various elements entering into machine design, such as cams, gears, sprocket-wheels, cone pulleys, bolts, screws, couplings, clutches, shafting and fly-wheels, have been treated in such a way as to make possible the use of the work as a text-book for a continuous course of study. It is easily comprehended and assimilated even by students of limited previous training. 330 pages, 215 engravings. Price **$2.00**

A New Sketching Paper.

A new specially ruled paper to enable you to make sketches or drawings in isometric perspective without any figuring or fussing. It is being used for shop details as well as for assembly drawings, as it makes one sketch do the work of three, and no workman can help seeing just what is wanted.

Pads of 40 sheets, 6x9 inches. Price . . . . **25 cents**
Pads of 40 sheets, 9x12 inches. Price . . . . **50 cents**
40 sheets, 12x18 inches. Price **$1.00**

ELECTRICITY

Arithmetic of Electricity. By Prof. T. O'Conor Sloane.

A practical treatise on electrical calculations of all kinds reduced to a series of rules, all of the simplest forms, and involving only ordinary arithmetic; each rule illustrated by one or more practical problems, with detailed solution of each one. This book is classed among the most useful works published on the science of electricity, covering as it does the mathematics of electricity in a manner that will attract the attention of those who are not familiar with algebraical formulas. 20th Edition. 160 pages. Price **$1.00**

Commutator Construction. By Wm. Baxter, Jr.

The business end of any dynamo or motor of the direct current type is the commutator. This book goes into the designing, building, and maintenance of commutators, shows how to locate troubles and how to remedy them; everyone who fusses with dynamos needs this. 4th Edition. Price . . . . **25 cents**

Dynamo Building for Amateurs, or How to Construct a Fifty-Watt Dynamo. By Arthur J. Weed, Member of N. Y. Electrical Society.

A practical treatise showing in detail the construction of a small dynamo or motor, the entire machine work of which can be done on a small foot lathe. Dimensioned working drawings are given for each piece of machine work, and each operation is clearly described. This machine, when used as a dynamo, has an output of fifty watts; when used as a motor it will drive a small drill press or lathe. It can be used to drive a sewing machine on any and all ordinary work. The book is illustrated with more than sixty original engravings, showing the actual construction of the different parts. Among the contents are chapters on: 1. Fifty-Watt Dynamo. 2. Side Bearing Rods. 3. Field Punching. 4 Bearings. 5. Commutator. 6. Pulley. 7. Brush Holders. 8 Connection Board. 9 Armature Shaft. 10. Armature. 11. Armature Winding. 12. Field Winding. 13. Connecting and starting.
Paper. Price . . . . **50 cents**
Cloth. Price **$1.00**

Electric Bells. By M. B. Sleeper.

A complete treatise for the practical worker in Installing, Operating and Testing Bell Circuits, Burglar Alarms, Thermostats, and other apparatus used with Electric Bells.
Both the electrician and the experimenter will find in this book new material which is essential in their work. Tools, bells, batteries, unusual circuits, burglar alarms, annunciator systems, thermostats, circuit breakers, time alarms, and other apparatus used in bell circuits are described from the standpoints of their application, construction and repair. The detailed instruction for building the apparatus will appeal to the experimenter particularly.
The practical worker will find the chapter on Wiring, Calculation of Wire Sizes and Magnet Winding, Upkeep of Systems, and the Location of Faults, of the greatest value in their work. Among the chapters are: Tools and Materials for Bell Work; How and Why Bell Work; Batteries for Small Installations; Making Bells and Push Buttons; Wiring Bell Systems; Construction of Annunciators and Signals; Burglary Alarms and Auxiliary Apparatus; More Elaborate Bell Systems; Finding Faults and Remedying Them. 124 pages, fully illustrated. Price . . . . **50 cents**

Electric Lighting and Heating Pocket Book. By Sydney F. Walker.

This book puts in convenient form useful information regarding the apparatus which is likely to be attached to the mains of an electrical company. Tables of units and equivalents are included and useful electrical laws and formulas are stated. 438 pages, 300 engravings. Bound in leather. Pocket book form. Price **$3.00**

Electric Wiring, Diagrams and Switchboards. By Newton Harrison, with additions by Thomas Poppe.

A thoroughly practical treatise covering the subject of Electric Wiring in all its branches, including explanations and diagrams which are thoroughly explicit and greatly simplify the subject. Practical, every-day problems in wiring are presented and the method of obtaining intelligent results clearly shown. Only arithmetic is used. Ohm's law is given a simple explanation with reference to wiring for direct and alternating currents. The fundamental principle of drop of potential in circuits is shown with its various applications. The simple circuit is developed with the position of mains, feeders and branches; their treatment as a part of a wiring plan and their employment in house wiring clearly illustrated. Some simple facts about testing are included in connection with the wiring. Molding and conduit work are given careful consideration; and switchboards are systematically treated, built up and illustrated, showing the purpose they serve, for connection with the circuits, and to shunt and compound wound machines. The simple principles of switchboard construction, the development of the switchboard, the connections of the various instruments, including the lightning arrester, are also plainly set forth.

Alternating current wiring is treated, with explanations of the power factor, conditions calling for various sizes of wire, and a simple way of obtaining the sizes for single-phase, two-phase and three-phase circuits. This is the only complete work issued showing and telling you what you should know about direct and alternating current wiring. It is a ready reference. The work is free from advanced technicalities and mathematics, arithmetic being used throughout. It is in every respect a handy, well-written, instructive, comprehensive volume on wiring for the wireman, foreman, contractor, or electrician. 2nd Revised Edition. 303 pages, 130 illustrations. Price **$1.50**

Electric Furnaces and their Industrial Applications. By J. WRIGHT.

This is a book which will prove of interest to many classes of people; the manufacturer who desires to know what product can be manufactured successfully in the electric furnace, the chemist who wishes to post himself on the electro-chemistry, and the student of science who merely looks into the subject from curiosity. New, Revised and Enlarged Edition. 320 pages. Fully illustrated, cloth. Price **$3.00**

Electric Toy Making, Dynamo Building, and Electric Motor Construction. By Prof. T. O'CONOR SLOANE.

This work treats of the making at home of electrical toys, electrical apparatus, motors, dynamos, and instruments in general, and is designed to bring within the reach of young and old the manufacture of genuine and useful electrical appliances. The work is especially designed for amateurs and young folks.

Thousands of our young people are daily experimenting, and busily engaged in making electrical toys and apparatus of various kinds. The present work is just what is wanted to give the much needed information in a plain, practical manner, with illustrations to make easy the carrying out of the work. 20th Edition. Price **$1.00**

Practical Electricity. By Prof. T. O'CONOR SLOANE.

This work of 768 pages was previously known as Sloane's Electricians' Hand Book, and is intended for the practical electrician who has to make things go. The entire field of electricity is covered within its pages. Among some of the subjects treated are: The Theory of the Electric Current and Circuit, Electro-Chemistry, Primary Batteries, Storage Batteries, Generation and Utilization of Electric Powers, Alternating Current, Armature Winding, Dynamos and Motors, Motor Generators, Operation of the Central Station Switchboards, Safety Appliances, Distribution of Electric Light and Power, Street Mains, Transformers, Arc and Incandescent Lighting, Electric Measurements, Photometry, Electric Railways, Telephony, Bell-Wiring, Electric-Plating, Electric Heating, Wireless Telegraphy, etc. It contains no useless theory; everything is to the point. It teaches you just what you want to know about electricity. It is the standard work published on the subject. Forty-one chapters, 556 engravings. Price **$2.50**

Electricity Simplified. By Prof. T. O'CONOR SLOANE.

The object of "Electricity Simplified" is to make the subject as plain as possible and to show what the modern conception of electricity is; to show how two plates of different metal, immersed in acid, can send a message around the globe; to explain how a bundle of copper wire rotated by a steam engine can be the agent in lighting our streets, to tell what the volt, ohm and ampere are, and what high and low tension mean; and to answer the questions that perpetually arise in the mind in this age of electricity. 13th Edition. 172 pages. Illustrated. Price **$1.00**

House Wiring. By THOMAS W. POPPE.

This work describes and illustrates the actual installation of Electric Light Wiring, the manner in which the work should be done, and the method of doing it. The book can be conveniently carried in the pocket. It is intended for the Electrician, Helper and Apprentice. It solves all Wiring Problems and contains nothing that conflicts with the rulings of the National Board of Fire Underwriters. It gives just the information essential to the Successful Wiring of a Building. Among the subjects treated are: Locating the Meter. Panel-Boards. Switches Plug Receptacles. Brackets. Ceiling Fixtures. The Meter Connections. The Feed Wires. The Steel Armored Cable System. The Flexible Steel Conduit System. The Ridig Conduit System. A digest of the National Board of Fire Underwriters' rules relating to metallic wiring systems. Various switching arrangements explained and diagrammed. The easiest method of testing the Three- and Four-way circuits explained. The grounding of all metallic wiring systems and the reason for doing so shown and explained. The insulation of the metal parts of lamp fixtures and the reason for the same described and illustrated. 125 pages. 2nd Edition, revised and enlarged. Fully illustrated. Flexible cloth. Price . . . . **50 cents**

How to Become a Successful Electrician. By Prof. T. O'Conor Sloane.

Every young man who wishes to become a successful electrician should read this book. It tells in simple language the surest and easiest way to become a successful electrician. The studies to be followed, methods of work, field of operation and the requirements of the successful electrician are pointed out and fully explained. Every young engineer will find this an excellent stepping stone to more advanced works on electricity which he must master before success can be attained. Many young men become discouraged at the very outstart by attempting to read and study books that are far beyond their comprehension. This book serves as the connecting link between the rudiments taught in the public schools and the real study of electricity. It is interesting from cover to cover. 18th Revised Edition, just issued. 205 pages. Illustrated. Price **$1.00**

Management of Dynamos. By Lummis-Paterson.

A handbook of theory and practice. This work is arranged in three parts. The first part covers the elementary theory of the dynamo The second part, the construction and action of the different classes of dynamos in common use are described; while the third part relates to such matters as affect the practical management and working of dynamos and motors. 4th Edition. 292 pages, 117 illustrations. Price **$1.50**

Standard Electrical Dictionary. By T. O'Conor Sloane.

An indispensable work to all interested in electrical science. Suitable alike for the student and professional. A practical handbook of reference containing definitions of about 5,000 distinct words, terms and phrases. The definitions are terse and concise and include every term used in electrical science. Recently issued. An entirely new edition Should be in the possession of all who desire to keep abreast with the progress of this branch of science. In its arrangement and typography the book is very convenient. The word or term defined is printed in black-faced type, which readily catches the eye, while the body of the page is in smaller but distinct type. The definitions are well worded, and so as to be understood by the non-technical reader. The general plan seems to be to give an exact, concise definition, and then amplify and explain in a more popular way. Synonyms are also given, and references to other words and phrases are made A very complete and accurate index of fifty pages is at the end of the volume; and as this index contains all synonyms, and as all phrases are indexed in every reasonable combination of words, reference to the proper place in the body of the book is readily made. It is difficult to decide how far a book of this character is to keep the dictionary form, and to what extent it may assume the encyclopedia form. For some purposes, concise, exactly worded definitions are needed; for other purposes, more extended descriptions are required. This book seeks to satisfy both demands, and does it with considerable success. 682 pages, 393 illustrations. 12th Edition.
Price **$3.00**

Storage Batteries Simplified. By Victor W. Pagé, M.E.

A complete treatise on storage battery operating principles, repairs and applications. The greatly increasing application of storage batteries in modern engineering and mechanical work has created a demand for a book that will consider this subject completely and exclusively. This is the most thorough and authoritative treatise ever published on this subject. It is written in easily understandable, non-technical language so that any one may grasp the basic principles of storage battery action as well as their practical industrial applications. All electric and gasoline automobiles use storage batteries. Every automobile repairman, dealer or salesman should have a good knowledge of maintenance and repair of these important elements of the motor car mechanism This book not only tells how to charge, care for and rebuild storage batteries but also outlines all the industrial uses. Learn how they run street cars, locomotives and factory trucks. Get an understanding of the important functions they perform in submarine boats, isolated lighting plants, railway switch and signal systems, marine applications, etc. This book tells how they are used in central station standby service, for starting automobile motors and in ignition systems Every practical use of the modern storage battery is outlined in this treatise. 320 pages, fully illustrated. Price . . . **$1.50**

Switchboards. By William Baxter, Jr.

This book appeals to every engineer and electrician who wants to know the practical side of things It takes up all sorts and conditions of dynamos, connections and circuits, and shows by diagram and illustration just how the switchboard should be connected. Includes direct and alternating current boards, also those for arc lighting, incandescent and power circuits Special treatment on high voltage boards for power transmission. 2nd Edition. 190 pages, Illustrated. Price **$1.50**

Telephone Construction, Installation, Wiring, Operation and Maintenance. By W. H. Radcliffe and H. C. Cushing.

This book is intended for the amateur, the wireman, or the engineer who desires to establish a means of telephonic communication between the rooms of his home, office, or shop. It deals only with such things as may be of use to him rather than with theories.
Gives the principles of construction and operation of both the Bell and Independent instruments; approved methods of installing and wiring them; the means of protecting them from lightning and abnormal currents; their connection together for operation as series or bridging stations; and rules for their inspection and maintenance. Line wiring and the wiring and operation of special telephone systems are also treated Intricate mathematics are avoided, and all apparatus, circuits and systems are thoroughly described. The appendix

contains definitions of units and terms used in the text. Selected wiring tables, which are very helpful, are also included. Among the subjects treated are Construction, Operation, and Installation of Telephone Instruments; Inspection and Maintenance of Telephone Instruments; Telephone Line Wiring; Testing Telephone Line Wires and Cables; Wiring and Operation of Special Telephone Systems, etc. 2nd Edition, Revised and Enlarged. 223 pages, 154 illustrations **$1.00**

Wireless Telegraphy and Telephony Simply Explained. By ALFRED P. MORGAN.

This is undoubtedly one of the most complete and comprehensible treatises on the subject ever published, and a close study of its pages will enable one to master all the details of the wireless transmission of messages. The author has filled a long-felt want and has succeeded in furnishing a lucid, comprehensible explanation in simple language of the theory and practice of wireless telegraphy and telephony.

Among the contents are: Introductory; Wireless Transmission and Reception—The Aerial System, Earth Connections—The Transmitting Apparatus, Spark Coils and Transformers, Condensers, Helixes, Spark Gaps, Anchor Gaps, Aerial Switches—The Receiving Apparatus, Detectors, etc.—Tuning and Coupling, Tuning Coils, Loose Couplers, Variable Condensers, Directive Wave Systems—Miscellaneous Apparatus, Telephone Receivers, Range of Stations, Static Interference—Wireless Telephones, Sound and Sound Waves, The Vocal Cords and Ear—Wireless Telephone, How Sounds Are Changed into Electric Waves—Wireless Telephones, The Apparatus—Summary. 154 pages, 156 engravings. Price **$1.00**

Wiring a House. By HERBERT PRATT.

Shows a house already built; tells just how to start about wiring it; where to begin; what wire to use; how to run it according to Insurance Rules; in fact, just the information you need. Directions apply equally to a shop. 4th Edition. Price . . . . **25 cents**

FACTORY MANAGEMENT, ETC.

Modern Machine Shop Construction, Equipment and Management. By O. E. PERRIGO, M.E.

The only work published that describes the modern machine shop or manufacturing plant from the time the grass is growing on the site intended for it until the finished product is shipped. By a careful study of its thirty-two chapters the practical man may economically build, efficiently equip, and successfully manage the modern machine shop or manufacturing establishment. Just the book needed by those contemplating the erection of modern shop buildings, the rebuilding and reorganization of old ones, or the introduction of modern shop methods, time and cost systems. It is a book written and illustrated by a practical shop man for practical shop men who are too busy to read *theories* and want *facts*. It is the most complete all-around book of its kind ever published. It is a practical book for practical men, from the apprentice in the shop to the president in the office. It minutely describes and illustrates the most simple and yet the most efficient time and cost system yet devised. 2nd Revised and Enlarged Edition, just issued. 384 pages, 219 illustrations. Price . . . **$5.00**

FUEL

Combustion of Coal and the Prevention of Smoke. By WM. M. BARR.

This book has been prepared with special reference to the generation of heat by the combustion of the common fuels found in the United States, and deals particularly with the conditions necessary to the economic and smokeless combustion of bituminous coals in Stationary and Locomotive Steam Boilers.

The presentation of this important subject is systematic and progressive. The arrangement of the book is in a series of practical questions to which are appended accurate answers, which describe in language, free from technicalities, the several processes involved in the furnace combustion of American fuels; it clearly states the essential requisites for perfect combustion, and points out the best methods for furnace construction for obtaining the greatest quantity of heat from any given quality of coal. Nearly 350 pages, fully illustrated. Price . . **$1.00**

Smoke Prevention and Fuel Economy. By BOOTH and KERSHAW.

A complete treatise for all interested in smoke prevention and combustion, being based on the German work of Ernst Schmatolla, but it is more than a mere translation of the German treatise, much being added. The authors show as briefly as possible the principles of fuel combustion, the methods which have been and are at present in use, as well as the proper scientific methods for obtaining all the energy in the coal and burning it without smoke. Considerable space is also given to the examination of the waste gases, and several of the representative English and American mechanical stoker and similar appliances are described. The losses carried away in the waste gases are thoroughly analyzed and discussed in the Appendix, and abstracts are also here given of various patents on combustion apparatus. The book is complete and contains much of value to all who have charge of large plants. 194 pages. Illustrated. Price **$2.50**

GAS ENGINES AND GAS

Gas, Gasoline and Oil Engines. By GARDNER D. HISCOX. Revised by VICTOR W. PAGÉ, M.E.

Just issued New 1918 Edition, Revised and Enlarged. Every user of a gas engine needs this book. Simple, instructive and right up-to-date. The only complete work on the subject. Tells all about internal combustion engineering, treating exhaustively on the design, construction and practical application of all forms of gas, gasoline, kerosene and crude petroleum-oil engines. Describes minutely all auxiliary systems, such as lubrication, carburetion and ignition. Considers the theory and management of all forms of explosive motors for stationary and marine work, automobiles, aeroplanes and motor-cycles. Includes also Producer Gas and Its Production. Invaluable instructions for all students, gas-engine owners, gas-engineers, patent experts, designers, mechanics, draftsmen and all having to do with the modern power. Illustrated by over 400 engravings, many specially made from engineering drawings, all in correct proportion. 650 pages, 435 engravings. Price . . . . **$2.50 net**

The Gasoline Engine on the Farm: Its Operation, Repair and Uses. By XENO W. PUTNAM.

This is a practical treatise on the Gasoline and Kerosene Engine intended for the man who wants to know just how to manage his engine and how to apply it to all kinds of farm work to the best advantage.
This book abounds with hints and helps for the farm and suggestions for the home and housewife. There is so much of value in this book that it is impossible to adequately describe it in such small space. Suffice to say that it is the kind of a book every farmer will appreciate and every farm home ought to have. Includes selecting the most suitable engine for farm work, its most convenient and efficient installation, with chapters on troubles, their remedies, and how to avoid them. The care and management of the farm tractor in plowing, harrowing, harvesting and road grading are fully covered; also plain directions are given for handling the tractor on the road. Special attention is given to relieving farm life of its drudgery by applying power to the disagreeable small tasks which must otherwise be done by hand. Many home-made contrivances for cutting wood, supplying kitchen, garden, and barn with water, loading, hauling and unloading hay, delivering grain to the bins or the feed trough are included; also full directions for making the engine milk the cows, churn, wash, sweep the house and clean the windows, etc. Very fully illustrated with drawings of working parts and cuts showing Stationary, Portable and Tractor Engines doing all kinds of farm work. All money-making farms utilize power. Learn how to utilize power by reading the pages of this book. It is an aid to the result getter, invaluable to the up-to-date farmer, student, blacksmith, implement dealer and, in fact, all who can apply practical knowledge of stationary gasoline engines or gas tractors to advantage. 530 pages. Nearly 180 engravings. Price **$2.00**

WHAT IS SAID OF THIS BOOK:

"Am much pleased with the book and find it to be very complete and up-to-date. I will heartily recommend it to students and farmers whom I think would stand in need of such a work, as I think it is an exceptionally good one "—*N. S. Gardiner*, Prof. in Charge, Clemson Agr. College of S. C.; Dept. of Agri. and Agri. Exp. Station, Clemson College, S. C.
"I feel that Mr. Putnam's book covers the main points which a farmer should know."—*R. T. Burdick*, Instructor in Agronomy, University of Vermont, Burlington, Vt.

Gasoline Engines: Their Operation, Use and Care. By A. HYATT VERRILL.

The simplest, latest and most comprehensive popular work published on Gasoline Engines, describing what the Gasoline Engine is; its construction and operation; how to install it; how to select it; how to use it and how to remedy troubles encountered. Intended for Owners, Operators and Users of Gasoline Motors of all kinds. This work fully describes and illustrates the various types of Gasoline Engines used in Motor Boats, Motor Vehicles and Stationary Work. The parts, accessories and appliances are described with chapters on ignition, fuel, lubrication, operation and engine troubles. Special attention is given to the care, operation and repair of motors, with useful hints and suggestions on emergency repairs and makeshifts. A complete glossary of technical terms and an alphabetically arranged table of troubles and their symptoms form most valuable and unique features of this manual. Nearly every illustration in the book is original, having been made by the author. Every page is full of interest and value. A book which you cannot afford to be without. 275 pages, 152 specially made engravings. Price **$1.50**

Gas Engine Construction, or How to Build a Half-horsepower Gas Engine. By PARSELL and WEED.

A practical treatise of 300 pages describing the theory and principles of the action of Gas Engines of various types and the design and construction of a half-horsepower Gas Engine, with illustrations of the work in actual progress, together with the dimensioned working drawings, giving clearly the sizes of the various details; for the student, the scientific investigator, and the amateur mechanic. This book treats of the subject more from the standpoint of practice than that of theory. The principles of operation of Gas Engines are clearly and simply described, and then the actual construction of a half-horsepower engine is taken up, step by step, showing in detail the making of the Gas Engine. 3rd Edition. 300 pages. Price **$2.50**

How to Run and Install Two- and Four-Cycle Marine Gasoline Engines. By C. Von Culin.

Revised and enlarged edition just issued   The object of this little book is to furnish a pocket instructor for the beginner, the busy man who uses an engine for pleasure or profit, but who does not have the time or inclination for a technical book, but simply to thoroughly understand how to properly operate, install and care for his own engine.   The index refers to each trouble, remedy, and subject alphabetically.   Being a quick reference to find the cause, remedy and prevention for troubles, and to become an expert with his own engine.   Pocket size. Paper binding.   Price . . . . **25 cents**

Modern Gas Engines and Producer Gas Plants. By R. E. Mathot.

A guide for the gas engine designer, user, and engineer in the construction, selection, purchase, installation, operation, and maintenance of gas engines   More than one book on gas engines has been written, but not one has thus far even encroached on the field covered by this book. Above all, Mr. Mathot's work is a practical guide.   Recognizing the need of a volume that would assist the gas engine user in understanding thoroughly the motor upon which he depends for power, the author has discussed his subject without the help of any mathematics and without elaborate theoretical explanations.   Every part of the gas engine is described in detail, tersely, clearly, with a thorough understanding of the requirements of the mechanic.   Helpful suggestions as to the purchase of an engine, its installation, care, and operation, form a most valuable feature of the work.   320 pages, 175 detailed illustrations.   Price . . . **$2.50**

The Modern Gas Tractor. By Victor W. Pagé, M.E.

A complete treatise describing all types and sizes of gasoline, kerosene and oil tractors   Considers design and construction exhaustively, gives complete instructions for care, operation and repair, outlines all practical applications on the road and in the field.   The best and latest work on farm tractors and tractor power plants.   A work needed by farmers, students, blacksmiths, mechanics, salesmen, implement dealers, designers and engineers.   2nd Edition, Revised.   504 pages, 228 illustrations, 3 folding plates.   Price **$2.00**

GEARING AND CAMS

Bevel Gear Tables. By D. Ag. Engstrom.

A book that will at once commend itself to mechanics and draftsmen.   Does away with all the trigonometry and fancy figuring on bevel gears, and makes it easy for anyone to lay them out or make them just right.   There are 36 full-page tables that show every necessary dimension for all sizes or combinations you're apt to need.   No puzzling, figuring or guessing.   Gives placing distance, all the angles (including cutting angles), and the correct cutter to use   A copy of this prepares you for anything in the bevel-gear line.   3rd Edition.   66 pages. Price **$1.00**

Change Gear Devices. By Oscar E. Perrigo.

A practical book for every designer, draftsman, and mechanic interested in the invention and development of the devices for feed changes on the different machines requiring such mechanism. All the necessary information on this subject is taken up, analyzed, classified, sifted, and concentrated for the use of busy men who have not the time to go through the masses of irrelevant matter with which such a subject is usually encumbered and select such information as will be useful to them.
It shows just what has been done, how it has been done, when it was done, and who did it. It saves time in hunting up patent records and re-inventing old ideas.   88 pages.   3rd Edition. Price **$1.00**

Drafting of Cams. By Louis Rouillion.

The laying out of cams is a serious problem unless you know how to go at it right.   This puts you on the right road for practically any kind of cam you are likely to run up against.   3rd Edition.   Price . . . . **25 cents**

HYDRAULICS

Hydraulic Engineering. By Gardner D. Hiscox.

A treatise on the properties, power, and resources of water for all purposes.   Including the measurement of streams, the flow of water in pipes or conduits; the horsepower of falling water, turbine and impact water-wheels, wave motors, centrifugal, reciprocating and air-lift pumps. With 300 figures and diagrams and 36 practical tables   All who are interested in water-works development will find this book a useful one, because it is an entirely practical treatise upon a subject of present importance and cannot fail in having a far-reaching influence, and for this reason should have a place in the working library of every engineer.   Among the subjects treated are: Historical Hydraulics; Properties of Water; Measurement of the Flow of Streams;

Flow from Sub-surface Orifices and Nozzles; Flow of Water in Pipes; Siphons of Various Kinds; Dams and Great Storage Reservoirs; City and Town Water Supply; Wells and Their Reinforcement; Air-lift Methods of Raising Water; Artesian Wells; Irrigation of Arid Districts; Water Power; Water Wheels; Pumps and Pumping Machinery; Reciprocating Pumps; Hydraulic Power Transmission; Hydraulic Mining; Canals; Ditches; Conduits and Pipe Lines; Marine Hydraulics; Tidal and Sea Wave Power, etc. 320 pages. Price . . . **$4.00**

ICE AND REFRIGERATION

Pocketbook of Refrigeration and Ice Making. By A. J. WALLIS-TAYLOR.

This is one of the latest and most comprehensive reference books published on the subject of refrigeration and cold storage. It explains the properties and refrigerating effect of the different fluids in use, the management of refrigerating machinery and the construction and insulation of cold rooms with their required pipe surface for different degrees of cold;. freezing mixtures and non-freezing brines, temperatures of cold rooms for all kinds of provisions, cold storage charges for all classes of goods, ice making and storage of ice, data and memoranda for constant reference by refrigerating engineers, with nearly one hundred tables containing valuable references to every fact and condition required in the installment and operation of a refrigerating plant. New edition just published. Price **$1.50**

INVENTIONS—PATENTS

Inventors' Manual: How to Make a Patent Pay.

This is a book designed as a guide to inventors in perfecting their inventions, taking out their patents and disposing of them. It is not in any sense a Patent Solicitor's Circular nor a Patent Broker's Advertisement. No advertisements of any description appear in the work. It is a book containing a quarter of a century's experience of a successful inventor, together with notes based upon the experience of many other inventors
Among the subjects treated in this work are: How to Invent. How to Secure a Good Patent. Value of Good Invention. How to Exhibit an Invention. How to Interest Capital. How to Estimate the Value of a Patent. Value of Design Patents. Value of Foreign Patents. Value of Small Inventions. Advice on Selling Patents. Advice on the Formation of Stock Companies. Advice on the Formation of Limited Liability Companies. Advice on Disposing of Old Patents. Advice as to Patent Attorneys. Advice as to Selling Agents. Forms of Assignments. License and Contracts. State Laws Concerning Patent Rights. 1900 Census of the United States by Counts of Over 10,000 Population. Revised Edition. 120 pages. Price **$1.00**

KNOTS

Knots, Splices and Rope Work. By A. HYATT VERRILL.

This is a practical book giving complete and simple directions for making all the most useful and ornamental knots in common use, with chapters on Splicing, Pointing, Seizing, Serving, etc. This book is fully illustrated with 154 original engravings, which show how each knot, tie or splice is formed, and its appearance when finished. The book will be found of the greatest value to Campers, Yachtsmen, Travelers, Boy Scouts, in fact, to anyone having occasion to use or handle rope or knots for any purpose. The book is thoroughly reliable and practical, and is not only a guide, but a teacher   It is the standard work on the subject. Among the contents are: 1. Cordage, Kinds of Rope. Construction of Rope, Parts of Rope Cable and Bolt Rope. Strength of Rope, Weight of Rope. 2. Simple Knots and Bends. Terms Used in Handling Rope. Seizing Rope. 3. Ties and Hitches. 4. Noose, Loops and Mooring Knots. 5. Shortenings, Grommets and Salvages. 6. Lashings, Seizings and Splices. 7. Fancy Knots and Rope Work. 128 pages, 150 original engravings. 2nd Revised Edition. Price . . . . **75 cents**

LATHE WORK

Lathe Design, Construction, and Operation, with Practical Examples of Lathe Work. By OSCAR E. PERRIGO.

A new, revised edition, and the only complete American work on the subject, written by a man who knows not only how work ought to be done, but who also knows how to do it, and how to convey this knowledge to others   It is strictly up-to-date in its descriptions and illustrations. Lathe history and the relations of the lathe to manufacturing are given; also a description of the various devices for feeds and thread-cutting mechanisms from early efforts in this direction to the present time. Lathe design is thoroughly discussed, including back gearing, driving cones, thread-cutting gears, and all the essential elements of the modern lathe. The classification of lathes is taken up, giving the essential differences of the several types of lathes including, as is usually understood, engine lathes, bench lathes, speed lathes, forge lathes, gap lathes, pulley lathes, forming lathes, multiple-spindle lathes, rapid-reduction lathes, precision lathes, turret lathes. special lathes, electrically driven lathes, etc. In addition to the complete exposition on construction and design, much practical matter on lathe installation, care and operation has been incorporated in the enlarged new edition All kinds of lathe attachments for drilling, milling, etc , are described and complete instructions are given to enable the novice machinist to grasp the art of lathe operation as well as the principles involved in design. A number of difficult machining operations are described at length and illustrated. The new edition has nearly 500 pages and 350 illustrations. Price. **$2.50**

WHAT IS SAID OF THIS BOOK:

"This is a lathe book from beginning to end, and is just the kind of a book which one delights to consult—a masterly treatment of the subject in hand."—*Engineering News*
"This work will be of exceptional interest to any one who is interested in lathe practice, as one very seldom sees such a complete treatise on a subject as this is on the lathe."—*Canadian Machinery.*

Practical Metal Turning. By JOSEPH G. HORNER.

A work of 404 pages, fully illustrated, covering in a comprehensive manner the modern practice of machining metal parts in the lathe, including the regular engine lathe, its essential design, its uses, its tools, its attachments, and the manner of holding the work and performing the operations The modernized engine lathe, its methods, tools and great range of accurate work. The turret lathe, its tools, accessories and methods of performing its functions. Chapters on special work, grinding, tool holders, speeds, feeds, modern tool steels. etc. Second edition **$3.50**

Turning and Boring Tapers. By FRED H. COLVIN.

There are two ways to turn tapers; the right way and one other. This treatise has to do with the right way; it tells you how to start the work properly, how to set the lathe, what tools to use and how to use them, and forty and one other little things that you should know. Fourth edition . . . . **25 cents**

LIQUID AIR

Liquid Air and the Liquefaction of Gases. By T. O'CONOR SLOANE.

This book gives the history of the theory, discovery and manufacture of Liquid Air, and contains an illustrated description of all the experiments that have excited the wonder of audiences all over the country. It shows how liquid air, like water, is carried hundreds of miles and is handled in open buckets. It tells what may be expected from it in the near future
A book that renders simple one of the most perplexing chemical problems of the century. Startling developments illustrated by actual experiments.
It is not only a work of scientific interest and authority, but is intended for the general reader, being written in a popular style—easily understood by every one. Second edition 365 pages Price. **$2.00**

LOCOMOTIVE ENGINEERING

Air-Brake Catechism. By ROBERT H. BLACKALL.

This book is a standard text-book. It covers the Westinghouse Air-Brake Equipment, including the No. 5 and the No 6 E -T. Locomotive Brake Equipment; the K (Quick Service) Triple Valve for Freight Service; and the Cross-Compound Pump. The operation of all parts of the apparatus is explained in detail, and a practical way of finding their peculiarities and defects, with a proper remedy, is given. It contains 2,000 questions with their answers, which will enable any railroad man to pass any examination on the subject of Air Brakes Endorsed and used by air-brake instructors and examiners on nearly every railroad in the United States Twenty-sixth edition. 411 pages, fully illustrated with colored plates and diagrams. Price. **$2.00**

American Compound Locomotives. By FRED H. COLVIN.

The only book on compounds for the engineman or shopman that shows in a plain, practical way the various features of compound locomotives in use. Shows how they are made, what to do when they break down or balk. Contains sections as follows: A Bit of History. Theory of Compounding Steam Cylinders. Baldwin Two-Cylinder Compound. Pittsburg Two-Cylinder Compound Rhose Island Compound. Richmond Compound Rogers Compound Schenectady Two-Cylinder Compound Vauclain Compound. Tandem Compounds. Baldwin Tandem The Colvin-Wightman Tandem Schenectady Tandem Balanced Locomotives. Baldwin Balanced Compound. Plans for Balancing. Locating Blows. Breakdowns Reducing Valves Drifting. Valve Motion Disconnecting. Power of Compound Locomotives Practical Notes
Fully illustrated and containing ten special "Duotone" inserts on heavy Plate Paper, showing different types of Compounds. 142 pages. Price. **$1.00**

Application of Highly Superheated Steam to Locomotives. By ROBERT GARBE.

A practical book which cannot be recommended too highly to those motive-power men who are anxious to maintain the highest efficiency in their locomotives. Contains special chapters on Generation of Highly Superheated Steam; Superheated Steam and the Two-Cylinder Simple Engine; Compounding and Superheating; Designs of Locomotive Superheaters; Constructive Details of Locomotives Using Highly Superheated Steam. Experimental and Working Results. Illustrated with folding plates and tables. Cloth. Price **$2.50**

Combustion of Coal and the Prevention of Smoke. By WM. M. BARR.

This book has been prepared with special reference to the generation of heat by the combustion of the common fuels found in the United States and deals particularly with the conditions necessary to the economic and smokeless combustion of bituminous coal in Stationary and Locomotive Steam Boilers.
Presentation of this important subject is systematic and progressive. The arrangement of the book is in a series of practical questions to which are appended accurate answers, which describe in language free from technicalities the several processes involved in the furnace combustion of American fuels; it clearly states the essential requisites for perfect combustion, and points out the best methods of furnace construction for obtaining the greatest quantity of heat from any given quality of coal. Nearly 350 pages, fully illustrated. Price **$1.00**

Diary of a Round-House Foreman. By T. S. REILLY.

This is the greatest book of railroad experiences ever published. Containing a fund of information and suggestions along the line of handling men, organizing, etc., that one cannot afford to miss. 176 pages. Price **$1.00**

Link Motions, Valves and Valve Setting. By FRED H. COLVIN, Associate Editor of "American Machinist."

A handy book for the engineer or machinist that clears up the mysteries of valve setting. Shows the different valve gears in use, how they work, and why. Piston and slide valves of different types are illustrated and explained. A book that every railroad man in the motive-power department ought to have. Contains chapters on Locomotive Link Motion, Valve Movements, Setting Slide Valves, Analysis by Diagrams, Modern Practice, Slip of Block, Slice Valves, Piston Valves, Setting Piston Valves, Joy-Allen Valve Gear, Walschaert Valve Gear, Gooch Valve Gear, Alfree-Hubbell Valve Gear, etc., etc. Fully illustrated. Price . . . . **50 cents**

Locomotive Boiler Construction. By FRANK A. KLEINHANS.

The construction of boilers in general is treated and, following this, the locomotive boiler is taken up in the order in which its various parts go through the shop. Shows all types of boilers used; gives details of construction; practical facts, such as life of riveting, punches and dies; work done per day, allowance for bending and flanging sheets and other data. Including the recent Locomotive Boiler Inspection Laws and Examination Questions with their answers for Government Inspectors. Contains chapters on Laying-Out Work; Flanging and Forging; Punching; Shearing; Plate Planing; General Tables; Finishing Parts; Bending; Machinery Parts; Riveting; Boiler Details; Smoke-Box Details, Assembling and Calking; Boiler-Shop Machinery, etc, etc
There isn't a man who has anything to do with boiler work, either new or repair work, who doesn't need this book. The manufacturer, superintendent, foreman and boiler worker— all need it. No matter what the type of boiler, you'll find a mint of information that you wouldn't be without. Over 400 pages, five large folding plates. Price **$3.00**

Locomotive Breakdowns and their Remedies. By GEO. L. FOWLER. Revised by WM. W. WOOD, Air-Brake Instructor. Just issued Revised pocket edition.

It is out of the question to try and tell you about every subject that is covered in this pocket edition of Locomotive Breakdowns. Just imagine all the common troubles that an engineer may expect to happen some time, and then add all of the unexpected ones, troubles that could occur, but that you have never thought about, and you will find that they are all treated with the very best methods of repair. Walschaert Locomotive Valve Gear Troubles, Electric Headlight Troubles, as well as Questions and Answers on the Air Brake are all included. 312 pages. 8th Revised Edition. Fully illustrated. Price **$1.00**

Locomotive Catechism. By ROBERT GRIMSHAW.

The revised edition of "Locomotive Catechism," by Robert Grimshaw, is a New Book from Cover to Cover. It contains twice as many pages and double the number of illustrations of previous editions. Includes the greatest amount of practical information ever published on the construction and management of modern locomotives. Specially Prepared Chapters on the Walschaert Locomotive Valve Gear, the Air-Brake Equipment and the Electric Headlight are given.

It commends itself at once to every Engineer and Fireman, and to all who are going in for examination or promotion. In plain language, with full, complete answers, not only all the questions asked by the examining engineer are given, but those which the young and less experienced would ask the veteran, and which old hands ask as "stickers." It is a veritable Encyclopedia of the Locomotive, is entirely free from mathematics, easily understood and thoroughly up to date Contains over 4,000 Examination Questions with their Answers. 825 pages, 437 illustrations, and 3 folding plates. 28th Revised Edition. Price..... $2.00

Practical Instructor and Reference Book for Locomotive Firemen and Engineers. By CHAS. F. LOCKHART.

An entirely new book on the Locomotive. It appeals to every railroad man, as it tells him how things are done and the right way to do them. Written by a man who has had years of practical experience in locomotive shops and on the road firing and running. The information given in this book cannot be found in any other similar treatise. Eight hundred and fifty-one questions with their answers are included, which will prove specially helpful to those preparing for examination. Practical information on: The Construction and Operation of Locomotives, Breakdowns and their Remedies, Air Brakes and Valve Gears. Rules and Signals are handled in a thorough manner. As a book of reference it cannot be excelled. The book is divided into six parts, as follows: 1. The Fireman's Duties. 2. General Description of the Locomotive. 3. Breakdowns and their Remedies. 4. Air Brakes. 5. Extracts from Standard Rules. 6 Questions for Examination. The 851 questions have been carefully selected and arranged. These cover the examinations required by the different railroads. 368 pages, 88 illustrations. Price $1.50

Prevention of Railroad Accidents, or Safety in Railroading. By GEORGE BRADSHAW.

This book is a heart-to-heart talk with Railroad Employees, dealing with facts, not theories, and showing the men in the ranks, from every-day experience, how accidents occur and how they may be avoided. The book is illustrated with seventy original photographs and drawings showing the safe and unsafe methods of work. No visionary schemes, no ideal pictures. Just Plain Facts and Practical Suggestions are given. Every railroad employee who reads the book is a better and safer man to have in railroad service. It gives just the information which will be the means of preventing many injuries and deaths. All railroad employees should procure a copy, read it, and do their part in preventing accidents. 169 pages. Pocket size. Fully illustrated. Price..50 cents

Train Rule Examinations Made Easy. By G. E. COLLINGWOOD.

This is the only practical work on train rules in print. Every detail is covered, and puzzling points are explained in simple, comprehensive language, making it a practical treatise for the Train Dispatcher, Engineman, Trainman, and all others who have to do with the movements of trains. Contains complete and reliable information of the Standard Code of Train Rules for single track. Shows Signals in Colors, as used on the different roads. Explains fully the practical application of train orders, giving a clear and definite understanding of all orders which may be used. The meaning and necessity for certain rules are explained in such a manner that the student may know beyond a doubt the rights conferred under any orders he may receive or the action required by certain rules. As nearly all roads require trainmen to pass regular examinations, a complete set of examination questions, with their answers, are included. These will enable the student to pass the required examinations with credit to himself and the road for which he works. 2nd Edition, Revised. 256 pages, fully illustrated, with Train Signals in Colors. Price. $1.25

The Walschaert and Other Modern Radial Valve Gears for Locomotives. By WM. W. WOOD.

If you would thoroughly understand the Walschaert Valve Gear you should possess a copy of this book, as the author takes the plainest form of a steam engine—a stationary engine in the rough, that will only turn its crank in one direction—and from it builds up, with the reader's help, a modern locomotive equipped with the Walschaert Valve Gear, complete. The points discussed are clearly illustrated: Two large folding plates that show the positions of the valves of both inside or outside admission type, as well as the links and other parts of the gear when the crank is at nine different points in its revolution, are especially valuable in making the movement clear. These employ sliding cardboard models which are contained in a pocket in the cover.
The book is divided into five general divisions, as follows: 1. Analysis of the gear. 2 Designing and erecting the gear. 3. Advantages of the gear. 4 Questions and answers relating to the Walschaert Valve Gear. 5. Setting valves with the Walschaert Valve Gear; the three primary types of locomotive valve motion; modern radial valve gears other than the Walschaert; the Hobart All-free Valve and Valve Gear, with questions and answers on breakdowns; the Baker-Pilliod Valve Gear; the Improved Baker-Pilliod Valve Gear, with questions and answers on breakdowns
The questions with full answers given will be especially valuable to firemen and engineers in preparing for an examination for promotion. 245 pages. 3rd Revised Edition. Price $1.50

Westinghouse E-T Air-Brake Instruction Pocket Book. By Wm. W. Wood, Air-Brake Instructor.

Here is a book for the railroad man, and the man who aims to be one. It is without doubt the only complete work published on the Westinghouse E-T Locomotive Brake Equipment. Written by an Air-Brake Instructor who knows just what is needed. It covers the subject thoroughly. Everything about the New Westinghouse Engine and Tender Brake Equipment, including the standard No. 5 and the Perfected No. 6 style of brake, is treated in detail. Written in plain English and profusely illustrated with Colored Plates, which enable one to trace the flow of pressures throughout the entire equipment. The best book ever published on the Air Brake. Equally good for the beginner and the advanced engineer. Will pass any one through any examination. It informs and enlightens you on every point. Indispensable to every engineman and trainman.
Contains examination questions and answers on the E-T equipment. Covering what the E-T Brake is. How it should be operated. What to do when defective. Not a question can be asked of the engineman up for promotion, on either the No. 5 or the No. 6 E-T equipment, that is not asked and answered in the book. If you want to thoroughly understand the E-T equipment get a copy of this book. It covers every detail. Makes Air-Brake troubles and examinations easy. Price.. **$1.50**

MACHINE-SHOP PRACTICE

American Tool Making and Interchangeable Manufacturing. By J. V. Woodworth.

A "shoppy" book, containing no theorizing, no problematical or experimental devices. There are no badly proportioned and impossible diagrams, no catalogue cuts, but a valuable collection of drawings and descriptions of devices, the rich fruits of the author's own experience. In its 500-odd pages the one subject only, Tool Making, and whatever relates thereto, is dealt with. The work stands without a rival. It is a complete, practical treatise, on the art of American Tool Making and system of interchangeable manufacturing as carried on to-day in the United States. In it are described and illustrated all of the different types and classes of small tools, fixtures, devices, and special appliances which are in general use in all machine-manufacturing and metal-working establishments where economy, capacity, and interchangeability in the production of machined metal parts are imperative. The science of jig making is exhaustively discussed, and particular attention is paid to drill jigs, boring, profiling and milling fixtures and other devices in which the parts to be machined are located and fastened within the contrivances. All of the tools, fixtures, and devices illustrated and described have been or are used for the actual production of work, such as parts of drill presses, lathes, patented machinery, typewriters, electrical apparatus, mechanical appliances, brass goods, composition parts, mould products, sheet-metal articles, drop-forgings, jewelry, watches, medals, coins, etc. 531 pages. Price.. **$4.00**

HENLEY'S ENCYCLOPEDIA OF PRACTICAL ENGINEERING AND ALLIED TRADES. Edited by Joseph G. Horner, A.M.I., M.E.

This set of five volumes contains about 2,500 pages with thousands of illustrations, including diagrammatic and sectional drawings with full explanatory details. This work covers the entire practice of Civil and Mechanical Engineering. The best known experts in all branches of engineering have contributed to these volumes. The Cyclopedia is admirably well adapted to the needs of the beginner and the self-taught practical man, as well as the mechanical engineer, designer, draftsman, shop superintendent, foreman, and machinist. The work will be found a means of advancement to any progressive man. It is encyclopedic in scope, thorough and practical in its treatment on technical subjects, simple and clear in its descriptive matter, and without unnecessary technicalities or formulæ. The articles are as brief as may be and yet give a reasonably clear and explicit statement of the subject, and are written by men who have had ample practical experience in the matters of which they write. It tells you all you want to know about engineering and tells it so simply, so concisely, that one cannot help but understand. As a work of reference it is without a peer. Complete set of five volumes, price **$25.00**

The Modern Machinist. By John T. Usher.

This is a book, showing by plain description and by profuse engravings made expressly for the work, all that is best, most advanced, and of the highest efficiency in modern machine-shop practice, tools and implements, showing the way by which and through which, as Mr. Maxim says, "American machinists have become and are the finest mechanics in the world." Indicating as it does, in every line, the familiarity of the author with every detail of daily experience in the shop, it cannot fail to be of service to any man practically connected with the shaping or finishing of metals.
There is nothing experimental or visionary about the book, all devices being in actual use and giving good results. It might be called a compendium of shop methods, showing a variety of special tools and appliances which will give new ideas to many mechanics, from the superintendent down to the man at the bench. It will be found a valuable addition to any machinist's library, and should be consulted whenever a new or difficult job is to be done, whether it is boring, milling, turning, or planing, as they are all treated in a practical manner. Fifth edition. 320 pages. 250 illustrations. Price.................... **$2.50**

THE WHOLE FIELD OF MECHANICAL MOVEMENTS COVERED BY MR. HISCOX'S TWO BOOKS

We publish two books by Gardner D. Hiscox that will keep you from "inventing" things that have been done before, and suggest ways of doing things that you have not thought of before   Many a man spends time and money pondering over some mechanical problem, only to learn, after he has solved the problem, that the same thing has been accomplished and put in practice by others long before.   Time and money spent in an effort to accomplish what has already been accomplished are time and money LOST.   The whole field of mechanics, every known mechanical movement, and practically every device are covered by these two books.   If the thing you want has been invented, it is illustrated in them.   If it hasn't been invented, then you'll find in them the nearest things to what you want, some movements or devices that will apply in your case, perhaps;   or which will give you a key from which to work.   No book or set of books ever published is of more real value to the Inventor, Draftsman, or practical Mechanic than the two volumes described below.

Mechanical Movements, Powers, and Devices. By GARDNER D. HISCOX.

This is a collection of 1,890 engravings of different mechanical motions and appliances, accompanied by appropriate text, making it a book of great value to the inventor, the draftsman. and to all readers with mechanical tastes.   The book is divided into eighteen sections or chapters, in which the subject-matter is classified under the following heads: Mechanical Powers;   Transmission of Power;   Measurement of Power;   Steam Power;   Air Power Appliances;   Electric Power and Construction;   Navigation and Roads;   Gearing;   Motion and Devices·   Controlling Motion;   Horological;   Mining;   Mill and Factory Appliances;   Construction and Devices;   Drafting Devices;   Miscellaneous Devices, etc.   15th Edition   400 octavo pages.   Price **$3.00**

Mechanical Appliances, Mechanical Movements and Novelties of Construction. By GARDNER D. HISCOX.

This is a supplementary volume to the one upon mechanical movements.   Unlike the first volume, which is more elementary in character, this volume contains illustrations and descriptions of many combinations of motions and of mechanical devices and appliances found in different lines of machinery, each device being shown by a line drawing with a description showing its working parts and the method of operation.   From the multitude of devices described and illustrated might be mentioned, in passing, such items as conveyors and elevators, Pony brakes, thermometers, various types of boilers, solar engines, oil-fuel burners, condensers, evaporators, Corliss and other valve gears, governors, gas engines, water motors of various descriptions, air ships, motors and dynamos, automobile and motor bicycles, railway lock signals, car couplers, link and gear motions, ball bearings, breech-block mechanism for heavy guns, and a large accumulation of others of equal importance.   One thousand specially made engravings.   396 octavo pages.   Fourth edition.   Price...................... **$3.00**

Machine-Shop Tools and Shop Practice. By W. H. VANDERVOORT.

A work of 555 pages and 673 illustrations, describing in every detail the construction, operation and manipulation of both hand and machine tools.   Includes chapters on filing, fitting and scraping surfaces; on drills, reamers, taps and dies; the lathe and its tools: planers, shapers, and their tools; milling machines and cutters; gear cutters and gear cutting; drilling machines and drill work; grinding machines and their work; hardening and tempering; gearing, belting and transmission machinery;   useful data and tables.   Sixth edition. Price... **$3.00**

Machine-Shop Arithmetic. By COLVIN-CHENEY.

This is an arithmetic of the things you have to do with daily.   It tells you plainly about: how to find areas in figures;   how to find surface or volume of balls or spheres;   handy ways for calculating;   about compound gearing;   cutting screw threads on any lathe;   drilling for taps;   speeds of drills;   taps, emery wheels, grindstones, milling cutters, etc ;   all about the Metric system with conversion tables;   properties of metals;   strength of bolts and nuts; decimal equivalent of an inch.   All sorts of machine-shop figuring and 1,001 other things, any one of which ought to be worth more than the price of this book to you, as it saves you the trouble of bothering the boss.   6th Edition.   131 pages.   Price . . . . . . **50 cents**

Modern Machine-Shop Construction, Equipment and Management. By OSCAR E. PERRIGO.

The only work published that describes the Modern Shop or Manufacturing Plant from the time the grass is growing on the site intended for it until the finished product is shipped   Just the book needed by those contemplating the erection of modern shop buildings, the rebuilding and reorganization of old ones, or the introduction of Modern Shop Methods, time and cost systems.   It is a book written and illustrated by a practical shop man for practical shop men who are too busy to read theories and want facts.   It is the most complete all-round book of its kind ever published.   Second Edition, Revised.   384 large quarto pages.   219 original and specially made illustrations.   2nd Revised and Enlarged Edition.   Price **$5.00**

Modern Milling Machines: Their Design, Construction, and Operation.
By Joseph G. Horner.

This book describes and illustrates the Milling Machine and its work in such a plain, clear and forceful manner, and illustrates the subject so clearly and completely, that the up-to-date machinist, student or mechanical engineer cannot afford to do without the valuable information which it contains. It describes not only the early machines of this class, but notes their gradual development into the splendid machines of the present day, giving the design and construction of the various types, forms, and special features produced by prominent manufacturers, American and foreign. 304 pages, 300 illustrations. Cloth. Price... **$4.00**

"Shop Kinks." By Robert Grimshaw.

A book of 400 pages and 222 illustrations, being entirely different from any other book on machine-shop practice. Departing from conventional style, the author avoids universal or common shop usage and limits his work to showing special ways of doing things better, more cheaply and more rapidly than usual. As a result the advanced methods of representative establishments of the world are placed at the disposal of the reader. This book shows the proprietor where large savings are possible, and how products may be improved To the employee it holds out suggestions that, properly applied, will hasten his advancement. No shop can afford to be without it. It bristles with valuable wrinkles and helpful suggestions. It will benefit all, from apprentice to proprietor. Every machinist, at any age, should study its pages. Fifth edition. Price... **$2.50**

Threads and Thread Cutting. By Colvin and Stabel.

This clears up many of the mysteries of thread-cutting, such as double and triple threads, internal threads, catching threads, use of hobs, etc. Contains a lot of useful hints and several tables. Third edition. Price... **25 cents**

MANUAL TRAINING

Economics of Manual Training. By Louis Rouillion.

The only book published that gives just the information needed by all interested in Manual Training, regarding Buildings, Equipment, and Supplies Shows exactly what is needed for all grades of the work from the Kindergarten to the High and Normal School. Gives itemized lists of everything used in Manual Training Work and tells just what it ought to cost. Also shows where to buy supplies, etc. Contains 174 pages, and is fully illustrated. Second edition. Price.. **$1.50**

MARINE ENGINEERING

The Naval Architect's and Shipbuilder's Pocket Book of Formulæ, Rules, and Tables and Marine Engineer's and Surveyor's Handy Book of Reference. By Clement Mackrow and Lloyd Woollard.

The eleventh Revised and Enlarged Edition of this most comprehensive work has just been issued It is absolutely indispensable to all engaged in the Shipbuilding Industry, as it condenses into a compact form all data and formulæ that are ordinarily required. The book is completely up to date, including among other subjects a section on Aeronautics. 750 pages, limp leather binding. Price..**$5.00 net**

Marine Engines and Boilers: Their Design and Construction. By Dr. G. Bauer, Leslie S. Robertson and S. Bryan Donkin.

In the words of Dr. Bauer, the present work owes its origin to an oft felt want of a condensed treatise embodying the theoretical and practical rules used in designing marine engines and boilers. The need of such a work has been felt by most engineers engaged in the construction and working of marine engines, not only by the younger men, but also by those of greater experience. The fact that the original German work was written by the chief engineer of the famous Vulcan Works, Stettin, is in itself a guarantee that this book is in all respects thoroughly up-to-date, and that it embodies all the information which is necessary for the design and construction of the highest types of marine engines and boilers. It may be said that the motive power which Dr Bauer has placed in the fast German liners that have been turned out of late years from the Stettin Works represent the very best practice in marine engineering of the present day. The work is clearly written, thoroughly systematic, theoretically sound; while the character of the plans, drawings, tables, and statistics is without reproach The illustrations are careful reproductions from actual working drawings, with some well-executed photographic views of completed engines and boilers. 744 pages, 550 illustrations and numerous tables. Cloth. Price.. **$9.00 net**

MINING

Ore Deposits, with a Chapter on Hints to Prospectors. By J. P. JOHNSON.

This book gives a condensed account of the ore deposits at present known in South Africa. It is also intended as a guide to the prospector. Only an elementary knowledge of geology and some mining experience are necessary in order to understand this work. With these qualifications, it will materially assist one in his search for metalliferous mineral occurrences and, so far as simple ores are concerned, should enable one to form some idea of the possibilities of any he may find. Illustrated. Cloth. Price **$2.00**

Practical Coal Mining. By T. H. COCKIN.

An important work, containing 428 pages and 213 illustrations, complete with practical details, which will intuitively impart to the reader not only a general knowledge of the principles of coal mining, but also considerable insight into allied subjects. The treatise is positively up-to-date in every instance, and should be in the hands of every colliery engineer, geologist, mine operator, superintendent, foreman, and all others who are interested in or connected with the industry. 3d Edition. Cloth. Price **$2.50**

Physics and Chemistry of Mining. By T. H. BYROM.

A practical work for the use of all preparing for examinations in mining or qualifying for colliery managers' certificates. The aim of the author in this excellent book is to place clearly before the reader useful and authoritative data which will render him valuable assistance in his studies. The only work of its kind published. The information incorporated in it will prove of the greatest practical utility to students, mining engineers, colliery managers, and all others who are specially interested in the present-day treatment of mining problems. 160 pages, illustrated. Price **$2.00**

PATTERN MAKING

Practical Pattern Making. By F. W. BARROWS.

This book, now in its second edition, is a comprehensive and entirely practical treatise on the subject of pattern making, illustrating pattern work in both wood and metal, and with definite instructions on the use of plaster of paris in the trade. It gives specific and detailed descriptions of the materials used by pattern makers, and describes the tools, both those for the bench and the more interesting machine tools, having complete chapters on the Lathe, the Circular Saw, and the Band Saw. It gives many examples of pattern work, each one fully illustrated and explained with much detail These examples, in their great variety, offer much that will be found of interest to all pattern makers, and especially to the younger ones, who are seeking information on the more advanced branches of their trade

In this second edition of the work will be found much that is new, even to those who have long practised this exacting trade. In the description of patterns as adapted to the Moulding Machine many difficulties which have long prevented the rapid and economical production of castings are overcome; and this great, new branch of the trade is given much space. Stripping plate and stool plate work and the less expensive vibrator, or rapping plate work, are all explained in detail.

Plain, every-day rules for lessening the cost of patterns, with a complete system of cost keeping, a detailed method of marking, applicable to all branches of the trade, with complete information showing what the pattern is, its specific title, its cost, date of production, material of which it is made, the number of pieces and core-boxes, and its location in the pattern safe, all condensed into a most complete card record, with cross index

The book closes with an original and practical method for the inventory and valuation of patterns. Containing nearly 350 pages and 170 illustrations. Price **$2.00**

PERFUMERY

Perfumes and Cosmetics: Their Preparation and Manufacture. By G. W. ASKINSON, Perfumer.

A comprehensive treatise, in which there has been nothing omitted that could be of value the perfumer or manufacturer of toilet preparations Complete directions for making hand or chief perfumes, smelling-salts, sachets, fumigating pastilles, preparations for the care of the skin, the mouth, the hair, cosmetics, hair dyes and other toilet articles are given, also a detailed description of aromatic substances; their nature, tests of purity, and wholesome manufacture, including a chapter on synthetic products, with formulas for their use. A book of general as well as professional interest, meeting the wants not only of the druggist and perfume manufacturer, but also of the general public Among the contents are: 1. The History of Perfumery 2 About Aromatic Substances in General 3 Odors from the Vegetable Kingdom 4 The Aromatic Vegetable Substances Employed in Perfumery. 5 The Animal Substances Used in Perfumery 6 The Chemical Products Used in Perfumery. 7 The Extraction of Odors 8 The Special Characteristics of Aromatic Substances 9 The Adulteration of Essential Oils and Their Recognition 10 Synthetic Products 11 Table of Physical Properties of Aromatic Chemicals 12 The Essences or Extracts Employed in Perfumery. 13 Directions for Making the Most Important Essences and Extracts.

14. The Division of Perfumery. 15. The Manufacture of Handkerchief Perfumes. 16. Formulas for Handkerchief Perfumes. 17. Ammoniacal and Acid Perfumes. 18. Dry Perfumes. 19 Formulas for Dry Perfumes. 20. The Perfumes Used for Fumigation 21. Antiseptic and Therapeutic Value of Perfumes 22. Classification of Odors 23 Some Special Perfumery Products. 24. Hygiene and Cosmetic Perfumery. 25. Preparations for the Care of the Skin. 26. Manufacture of Casein. 27. Formulas for Emulsions 28 Formulas for Cream. 29. Formulas for Meals, Pastes and Vegetable Milk. 30. Preparations Used for the Hair. 31. Formulas for Hair Tonics and Restorers 32. Pomades and Hair Oils. 33. Formulas for the Manufacture of Pomades and Hair Oils. 34. Hair Dyes and Depilatories. 35. Wax Pomades, Bandolines and Brilliantines 36. Skin Cosmetics and Face Lotions. 37. Preparations for the Nails. 38. Water Softeners and Bath Salts. 39. Preparations for the Care of the Mouth 40. The Colors Used in Perfumery. 41. The Utensils Used in the Toilet. Fourth edition, much enlarged and brought up to date. Nearly 400 pages, illustrated. Price **$5.00**

WHAT IS SAID OF THIS BOOK:

"The most satisfactory work on the subject of Perfumery that we have ever seen."
"We feel safe in saying that here is a book on Perfumery that will not disappoint you, for it has practical and excellent formulæ that are within your ability to prepare readily."
"We recommend the volume as worthy of confidence, and say that no purchaser will be disappointed in securing from its pages good value for its cost, and a large dividend on the same, even if he should use but one per cent. of its working formulæ. There is money in it for every user of its information."—*Pharmaceutical Record.*

PLUMBING

Mechanical Drawing for Plumbers. By R. M. STARBUCK.

A concise, comprehensive and practical treatise on the subject of mechanical drawing in its various modern applications to the work of all who are in any way connected with the plumbing trade. Nothing will so help the plumber in estimating and in explaining work to customers and workmen as a knowledge of drawing, and to the workman it is of inestimable value if he is to rise above his position to positions of greater responsibility. Among the chapters contained are: 1. Value to plumber of knowledge of drawing; tools required and their use; common views needed in mechanical drawing. 2. Perspective versus mechanical drawing in showing plumbing construction. 3. Correct and incorrect methods in plumbing drawing; plan and elevation explained. 4. Floor and cellar plans and elevation; scale drawings; use of triangles. 5. Use of triangles; drawing of fittings, traps, etc. 6. Drawing plumbing elevations and fittings 7 Instructions in drawing plumbing elevations. 8. The drawing of plumbing fixtures; scale drawings. 9 Drawings of fixtures and fittings. 10. Inking of drawings 11. Shading of drawings. 12. Shading of drawings. 13. Sectional drawings; drawing of threads. 14. Plumbing elevations from architect's plan. 15 Elevations of separate parts of the plumbing system 16. Elevations from the architect's plans. 17 Drawings of detail plumbing connections 18. Architect's plans and plumbing elevations of residence. 19. Plumbing elevations of residence (*continued*); plumbing plans for cottage. 20. Plumbing elevations; roof connections. 21. Plans and plumbing elevations for six-flat building. 22. Drawing of various parts of the plumbing system; use of scales. 23. Use of architect's scales. 24. Special features in the illustrations of country plumbing. 25. Drawing of wrought-iron piping, valves, radiators, coils, etc. 26. Drawing of piping to illustrate heating systems. 150 illustrations. Price **$1.50**

Modern Plumbing Illustrated. By R. M. STARBUCK.

This book represents the highest standard of plumbing work. It has been adopted and used as a reference book by the United States Government in its sanitary work in Cuba, Porto Rico and the Philippines, and by the principal Boards of Health of the United States and Canada.

It gives connections, sizes and working data for all fixtures and groups of fixtures. It is helpful to the master plumber in demonstrating to his customers and in figuring work. It gives the mechanic and student quick and easy access to the best modern plumbing practice. Suggestions for estimating plumbing construction are contained in its pages. This book represents, in a word, the latest and best up-to-date practice and should be in the hands of every architect, sanitary engineer and plumber who wishes to keep himself up to the minute on this important feature of construction. Contains following chapters, each illustrated with a full-page plate: Kitchen sink, laundry tubs, vegetable wash sink; lavatories, pantry sinks, contents of marble slabs; bath tub, foot and sitz bath, shower bath; water closets. venting of water closets; low-down water closets, water closets operated by flush valves, water closet range; slop sink, urinals, the bidet; hotel and restaurant sink, grease trap; refrigerators, safe wastes, laundry waste, lines of refrigerators, bar sinks, soda fountain sinks; horse stall, frost-proof water closets; connections for S traps, venting; connections for drum traps; soil-pipe connections; supporting of soil pipe; main trap and fresh-air inlet; floor drains and cellar drains, subsoil drainage; water closets and floor connections; local venting; connections for bath rooms; connections for bath rooms, *continued;* examples of poor practice; roughing work ready for test; testing of plumbing systems; method of continuous venting; continuous venting for two-floor work; continuous venting for two lines of fixtures on three or more floors; continuous venting of water closets; plumbing for cottage house; construction for cellar piping; plumbing for residence, use of special fittings; plumbing for two-flat house; plumbing for apartment building, plumbing for double apartment building; plumbing for office building; plumbing for public toilet rooms; plumbing for public toilet rooms, *continued;* plumbing for bath establishment; plumbing for engine house, factory plumbing; automatic flushing for schools, factories, etc.; use of flushing valves; urinals for public toilet rooms; the Durham system, the destruction of pipes by electrolysis; construction of work without use of lead; automatic sewage lift; automatic sump tank; country plumbing; construction of cesspools; septic tank and automatic sewage siphon; water supply for country house; thawing of water mains and service by electricity; double boilers; hot water supply of large buildings; automatic control of hot-water tank, suggestions for estimating plumbing construction. 407 octavo pages, fully illustrated by 57 full-page engravings. Third, revised and enlarged edition, just issued. Price **$4.00**

Standard Practical Plumbing. By R. M. STARBUCK.

A complete practical treatise of 450 pages, covering the subject of Modern Plumbing in all its branches, a large amount of space being devoted to a very complete and practical treatment of the subject of Hot Water Supply and Circulation and Range Boiler Work. Its thirty chapters include about every phase of the subject one can think of, making it an indispensable work to the master plumber, the journeyman plumber, and the apprentice plumber, containing chapters on: the plumber's tools; wiping solder; composition and use; joint wiping; lead work; traps; siphonage of traps; venting; continuous venting; house sewer and sewer connections; house drain; soil piping, roughing; main trap and fresh air inlet; floor, yard, cellar drains, rain leaders, etc ; fixture wastes; water closets; ventilation; improved plumbing connections; residence plumbing; plumbing for hotels, schools, factories, stables, etc.; modern country plumbing; filtration of sewage and water supply; hot and cold supply; range boilers; circulation; circulating pipes; range boiler problems; hot water for large buildings; water lift and its use; multiple connections for hot water boilers; heating of radiation by supply system; theory for the plumber; drawing for the plumber. Fully illustrated by 347 engravings. Price **$3.00**

RECIPE BOOK

Henley's Twentieth Century Book of Recipes, Formulas and Processes. Edited by GARDNER D. HISCOX.

The most valuable Techno-chemical Formula Book published, including over 10,000 selected scientific, chemical, technological, and practical recipes and processes
This is the most complete Book of Formulas ever published, giving thousands of recipes for the manufacture of valuable articles for everyday use. Hints, Helps, Practical Ideas, and Secret Processes are revealed within its pages. It covers every branch of the useful arts and tells thousands of ways of making money, and is just the book everyone should have at his command.
Modern in its treatment of every subject that properly falls within its scope, the book may truthfully be said to present the very latest formulas to be found in the arts and industries, and to retain those processes which long experience has proven worthy of a permanent record. To present here even a limited number of the subjects which find a place in this valuable work would be difficult. Suffice to say that in its pages will be found matter of intense interest and immeasurably practical value to the scientific amateur and to him who wishes to obtain a knowledge of the many processes used in the arts, trades and manufacture, a knowledge which will render his pursuits more instructive and remunerative. Serving as a reference book to the small and large manufacturer and supplying intelligent seekers with the information necessary to conduct a process, the work will be found of inestimable worth to the Metallurgist, the Photographer, the Perfumer, the Painter, the Manufacturer of Glues, Pastes, Cements, and Mucilages, the Compounder of Alloys, the Cook, the Physician, the Druggist, the Electrician, the Brewer, the Engineer, the Foundryman, the Machinist, the Potter, the Tanner, the Confectioner, the Chiropodist, the Manicurist, the Manufacturer of Chemical Novelties and Toilet Preparations, the Dyer, the Electroplater, the Enameler, the Engraver, the Provisioner, the Class Worker, the Goldbeater, the Watchmaker, the Jeweler, the Hat Maker, the Ink Manufacturer, the Optician, the Farmer, the Dairyman, the Paper Maker, the Wood and Metal Worker, the Chandler and Soap Maker, the Veterinary Surgeon, and the Technologist in general.
A mine of information, and up-to-date in every respect. A book which will prove of value to EVERYONE, as it covers every branch of the Useful Arts. Every home needs this book; every office, every factory, every store, every public and private enterprise—EVERYWHERE —should have a copy. 800 pages. Price **$3.00**

WHAT IS SAID OF THIS BOOK:

"Your Twentieth Century Book of Recipes, Formulas, and Processes duly received. I am glad to have a copy of it, and if I could not replace it, money couldn't buy it. It is the best thing of the sort I ever saw." (Signed) M. E. TRUX, Sparta, Wis.

"There are few persons who would not be able to find in the book some single formula that would repay several times the cost of the book."—*Merchants' Record and Show Window.*

"I purchased your book, 'Henley's Twentieth Century Book of Recipes, Formulas and Processes,' about a year ago and it is worth its weight in *gold.*"—WM. H. MURRAY, Bennington, Vt.

"ONE OF THE WORLD'S MOST USEFUL BOOKS"

"Some time ago I got one of your 'Twentieth Century Books of Formulas,' and have made my living from it ever since. I am alone since my husband's death with two small children to care for and am trying so hard to support them. I have customers who take from me Toilet Articles I put up, following directions given in the book, and I have found everyone of them to be fine."—MRS. J. H. McMAKEN, West Toledo, Ohio.

RUBBER

Rubber Hand Stamps and the Manipulation of India Rubber. By T. O'Conor Sloane.

This book gives full details on all points, treating in a concise and simple manner the elements of nearly everything it is necessary to understand for a commencement in any branch of the India Rubber Manufacture. The making of all kinds of Rubber Hand Stamps, Small Articles of India Rubber, U. S. Government Composition, Dating Hand Stamps, the Manipulation of Sheet Rubber, Toy Balloons, India Rubber Solutions, Cements, Blackings, Renovating, Varnish, and Treatment for India Rubber Shoes, etc.; the Hektograph Stamp Inks, and Miscellaneous Notes, with a Short Account of the Discovery, Collection and Manufacture of India Rubber, are set forth in a manner designed to be readily understood, the explanations being plain and simple. Including a chapter on Rubber Tire Making and Vulcanizing; also a chapter on the uses of rubber in Surgery and Dentistry. 3rd Revised and Enlarged Edition. 175 pages. Illustrated **$1.00**

SAWS

Saw Filing and Management of Saws. By Robert Grimshaw.

A practical hand-book on filing, gumming, swaging, hammering, and the brazing of band saws, the speed, work, and power to run circular saws, etc. A handy book for those who have charge of saws, or for those mechanics who do their own filing, as it deals with the proper shape and pitches of saw teeth of all kinds and gives many useful hints and rules for gumming, setting, and filing, and is a practical aid to those who use saws for any purpose. Complete tables of proper shape, pitch, and saw teeth as well as sizes and number of teeth of various saws are included. 3rd Edition, Revised and Enlarged. Illustrated. Price. **$1.00**

STEAM ENGINEERING

American Stationary Engineering. By W. E. Crane.

This book begins at the boiler room and takes in the whole power plant. A plain talk on every-day work about engines, boilers, and their accessories. It is not intended to be scientific or mathematical. All formulas are in simple form so that any one understanding plain arithmetic can readily understand any of them. The author has made this the most practical book in print; has given the results of his years of experience, and has included about all that has to do with an engine room or a power plant. You are not left to guess at a single point. You are shown clearly what to expect under the various conditions; how to secure the best results; ways of preventing "shut downs" and repairs; in short, all that goes to make up the requirements of a good engineer, capable of taking charge of a plant. It's plain enough for practical men and yet of value to those high in the profession.
A partial list of contents is: The boiler room, cleaning boilers, firing, feeding; pumps, inspection and repair; chimneys, sizes and cost; piping; mason work; foundations; testing cement; pile driving; engines, slow and high speed; valves; valve setting; Corliss engines, setting valves, single and double eccentric; air pumps and condensers; different types of condensers; water needed; lining up; pounds; pins not square in crosshead or crank; engineers' tools; pistons and piston rings; bearing metal; hardened copper; drip pipes from cylinder jacket; belts, how made, care of; oils; greases; testing lubricants; rules and tables, including steam tables; areas of segments; squares and square roots; cubes and cube root; areas and circumferences of circles. Notes on: Brick work; explosions; pumps; pump valves; heaters, economizers; safety valves; lap, lead, and clearance. Has a complete examination for a license, etc., etc. 3rd Edition. 345 pages, illustrated. Price . . . **$2.00**

Engine Runner's Catechism. By Robert Grimshaw.

A practical treatise for the stationary engineer, telling how to erect, adjust, and run the principal steam engines in use in the United States Describing the principal features of various special and well-known makes of engines: Temper Cut-off, Shipping and Receiving Foundations, Erecting and Starting, Valve Setting, Care and Use, Emergencies, Erecting and Adjusting Special Engines.
The questions asked throughout the catechism are plain and to the point, and the answers are given in such simple language as to be readily understood by anyone. All the instructions given are complete and up-to-date; and they are written in a popular style, without any technicalities or mathematical formulæ. The work is of a handy size for the pocket, clearly and well printed, nicely bound, and profusely illustrated.
To young engineers this catechism will be of great value, especially to those who may be preparing to go forward to be examined for certificates of competency; and to engineers generally it will be of no little service, as they will find in this volume more really practical and useful information than is to be found anywhere else within a like compass. 387 pages. 7th Edition Price **$2.00**

Modern Steam Engineering in Theory and Practice. By GARDNER D. HISCOX.

This is a complete and practical work issued for Stationary Engineers and Firemen, dealing with the care and management of boilers, engines, pumps, superheated steam, refrigerating machinery, dynamos, motors, elevators, air compressors, and all other branches with which the modern engineer must be familiar. Nearly 200 questions with their answers on steam and electrical engineering, likely to be asked by the Examining Board, are included. Among the chapters are: Historical: steam and its properties; appliances for the generation of steam; types of boilers; chimney and its work; heat economy of the feed water; steam pumps and their work; incrustation and its work; steam above atmospheric pressure; flow of steam from nozzles; superheated steam and its work; adiabatic expansion of steam; indicator and its work; steam engine proportions; slide valve engines and valve motion; Corliss engine and its valve gear; compound engine and its theory; triple and multiple expansion engine; steam turbine; refrigeration; elevators and their management; cost of power; steam engine troubles; electric power and electric plants. 487 pages, 405 engravings. 3rd Edition. Price . . . . : **$3.00**

Steam Engine Catechism. By ROBERT GRIMSHAW.

This unique volume of 413 pages is not only a catechism on the question and answer principle but it contains formulas and worked-out answers for all the Steam problems that appertain to operation and management of the Steam Engine. Illustrations of various valves and valve gear with their principles of operation are given. Thirty-four Tables that are indispensable to every engineer and fireman that wishes to be progressive and is ambitious to become master of his calling are within its pages. It is a most valuable instructor in the service of Steam Engineering. Leading engineers have recommended it as a valuable educator for the beginner as well as a reference book for the engineer. It is thoroughly indexed for every detail. Every essential question on the Steam Engine with its answer is contained in this valuable work. 16th Edition. Price **$2.00**

Steam Engineer's Arithmetic. By COLVIN-CHENEY.

A practical pocket-book for the steam engineer. Shows how to work the problems of the engine room and shows "why" Tells how to figure horsepower of engines and boilers; area of boilers; has tables of areas and circumferences; steam tables; has a dictionary of engineering terms Puts you on to all of the little kinks in figuring whatever there is to figure around a power plant Tells you about the heat unit; absolute zero; adiabatic expansion; duty of engines; factor of safety; and a thousand and one other things; and everything is plain and simple—not the hardest way to figure, but the easiest. 2nd Edition. Price . . **50 Cents**

Engine Tests and Boiler Efficiencies. By J. BUCHETTI.

This work fully describes and illustrates the method of testing the power of steam engines, turbines and explosive motors The properties of steam and the evaporative power of fuels. Combustion of fuel and chimney draft; with formulas explained or practically computed. 255 pages, 179 illustrations. Price **$3.00**

Horsepower · Chart.

Shows the horsepower of any stationary engine without calculation. No matter what the cylinder diameter of stroke, the steam pressure of cut-off, the revolutions, or whether condensing or non-condensing, it's all there. Easy to use, accurate, and saves time and calculations. Especially useful to engineers and designers. Price . . . . **50 Cents**

STEAM HEATING AND VENTILATION

Practical Steam, Hot-Water Heating and Ventilation. By A. G. KING.

This book is the standard and latest work published on the subject and has been prepared for the use of all engaged in the business of steam, hot-water heating, and ventilation It is an original and exhaustive work. Tells how to get heating contracts, how to install heating and ventilating apparatus, the best business methods to be used, with "Tricks of the Trade" for shop use. Rules and data for estimating radiation and cost and such tables and information as make it an indispensable work for everyone interested in steam, hot-water heating, and ventilation. It describes all the principal systems of steam, hot-water, vacuum, vapor, and vacuum-vapor heating, together with the new accelerated systems of hot-water circulation, including chapters on up-to-date methods of ventilation and the fan or blower system of heating and ventilation. Containing chapters on: I. Introduction. II. Heat III Evolution of artificial heating apparatus IV. Boiler surface and settings V. The chimney flue. VI. Pipe and fittings. VII. Valves, various kinds. VIII. Forms of radiating surfaces IX.

Locating of radiating surfaces. X. Estimating radiation. XI. Steam-heating apparatus. XII. Exhaust-steam heating. XIII. Hot-water heating. XIV. Pressure systems of hot-water work. XV. Hot-water appliances. XVI. Greenhouse heating. XVII. Vacuum vapor and vacuum exhaust heating. XVIII. Miscellaneous heating. XIX. Radiator and pipe connections. XX. Ventilation. XXI. Mechanical ventilation and hot-blast heating. XXII. Steam appliances. XXIII. District heating. XXIV Pipe and boiler covering. XXV. Temperature regulation and heat control. XXVI. Business methods. XXVII. Miscellaneous. XXVIII. Rules, tables, and useful information. 367 pages, 300 detailed engravings   2nd Edition—Revised. Price **$3.00**

Five Hundred Plain Answers to Direct Questions on Steam, Hot-Water, Vapor and Vacuum Heating Practice. By ALFRED G. KING.

This work, just off the press, is arranged in question and answer form; it is intended as a guide and text-book for the younger, inexperienced fitter and as a reference book for all fitters. This book tells "how" and also tells "why". No work of its kind has ever been published. It answers all the questions regarding each method or system that would be asked by the steam fitter or heating contractor, and may be used as a text or reference book, and for examination questions by Trade Schools or Steam Fitters' Associations. Rules, data, tables and descriptive methods are given, together with much other detailed information of daily practical use to those engaged in or interested in the various methods of heating. Valuable to those preparing for examinations. Answers every question asked relating to modern Steam, Hot-Water, Vapor and Vacuum Heating. Among the contents are: The Theory and Laws of Heat. Methods of Heating. Chimneys and Flues. Boilers for Heating. Boiler Trimmings and Settings. Radiation. Steam Heating. Boiler, Radiator and Pipe Connections for Steam Heating. Hot Water Heating. The Two-Pipe Gravity System of Hot Water Heating. The Circuit System of Hot Water Heating. The Overhead System of Hot Water Heating. Boiler, Radiator and Pipe Connections for Gravity Systems of Hot Water Heating. Accelerated Hot Water Heating. Expansion Tank Connections. Domestic Hot Water Heating. Valves and Air Valves. Vacuum Vapor and Vacuo-Vapor Heating. Mechanical Systems of Vacuum Heating. Non-Mechanical Vacuum Systems. Vapor Systems Atmospheric and Modulating Systems. Heating Greenhouses. Information, Rules and Tables. 200 pages, 127 illustrations. Octavo. Cloth. Price **$1.50**

STEEL

Steel: Its Selection, Annealing, Hardening, and Tempering. By E. R. MARKHAM.

This work was formerly known as "The American Steel Worker," but on the publication of the new, revised edition, the publishers deemed it advisable to change its title to a more suitable one. It is the standard work on Hardening, Tempering, and Annealing Steel of all kinds. This book tells how to select, and how to work, temper, harden, and anneal steel for everything on earth. It doesn't tell how to temper one class of tools and then leave the treatment of another kind of tool to your imagination and judgment, but it gives careful instructions for every detail of every tool, whether it be a tap, a reamer or just a screw-driver. It tells about the tempering of small watch springs, the hardening of cutlery, and the annealing of dies. In fact, there isn't a thing that a steel worker would want to know that isn't included. It is the standard book on selecting, hardening and tempering all grades of steel. Among the chapter headings might be mentioned the following subjects: Introduction; the workman; steel; methods of heating; heating tool steel; forging; annealing; hardening baths; baths for hardening; hardening steel; drawing the temper after hardening; examples of hardening; pack hardening; case hardening; spring tempering; making tools of machine steel; special steels; steel for various tools; causes of trouble; high-speed steels, etc. 400 pages. Very fully illustrated. Fourth edition. Price **$2.50**

Hardening, Tempering, Annealing, and Forging of Steel. By J. V. WOODWORTH.

A new work treating in a clear, concise manner all modern processes for the heating, annealing, forging, welding, hardening and tempering of steel, making it a book of great practical value to the metal-working mechanic in general, with special directions for the successful hardening and tempering of all steel tools used in the arts, including milling cutters, taps, thread dies, reamers, both solid and shell, hollow mills, punches and dies, and all kinds of sheet-metal working tools, shear blades, saws, fine cutlery, and metal-cutting tools of all description, as well as for all implements of steel both large and small. In this work the simplest and most satisfactory hardening and tempering processes are given.
The uses to which the leading brands of steel may be adapted are concisely presented, and their treatment for working under different conditions explained, also the special methods for the hardening and tempering of special brands
A chapter devoted to the different processes for case-hardening is also included, and special reference made to the adaptation of machinery steel for tools of various kinds. Fourth edition. 288 pages. 201 illustrations. Price **$2.50**

TRACTORS

The Modern Gas Tractor. By Victor W. Pagé, M.E.

A complete treatise describing all types and sizes of gasoline, kerosene and oil tractors Considers design and construction exhaustively, gives complete instructions for care, operation and repair, outlines all practical applications on the road and in the field. The best and latest work on farm tractors and tractor power plants. A work needed by farmers, students, blacksmiths, mechanics, salesmen, implement dealers, designers, and engineers Second edition, revised and enlarged. 504 pages. Nearly 300 illustrations and folding plates. Price **$2.00**

TURBINES

Marine Steam Turbines. By Dr. G. Bauer and O. Lasche. Assisted by E. Ludwig and H. Vogel.

Translated from the German and edited by M. G. S. Swallow. The book is essentially practical and discusses turbines in which the full expansion of steam passes through a number of separate turbines arranged for driving two or more shafts, as in the Parsons system, and turbines in which the complete expansion of steam from inlet to exhaust pressure occurs in a turbine on one shaft, as in the case of the Curtis machines. It will enable a designer to carry out all the ordinary calculation necessary for the construction of steam turbines, hence it fills a want which is hardly met by larger and more theoretical works. Numerous tables, curves and diagrams will be found, which explain with remarkable lucidity the reason why turbine blades are designed as they are, the course which steam takes through turbines of various types, the thermodynamics of steam turbine calculation, the influence of vacuum on steam consumption of steam turbines, etc In a word, the very information which a designer and builder of steam turbines most requires Large octavo, 214 pages. Fully illustrated and containing eighteen tables, including an entropy chart. Price, net **$3.50**

WATCH MAKING

Watchmaker's Handbook. By Claudius Saunier.

No work issued can compare with this book for clearness and completeness. It contains 498 pages and is intended as a workshop companion for those engaged in watch-making and allied mechanical arts. Nearly 250 engravings and fourteen plates are included. This is the standard work on watch-making. Price **$3.00**

WELDING

Automobile Welding with the Oxy-Acetylene Flame. By M. Keith Dunham.

Explains in a simple manner apparatus to be used, its care, and how to construct necessary shop equipment Proceeds then to the actual welding of all automobile parts, in a manner understandable by every one. *Gives principles never to be forgotten.* Aluminum, cast iron, steel, copper, brass, bronze, and malleable iron are fully treated, as well as a clear explanation of the proper manner to burn the carbon out of the combustion head. This book is of utmost value, since the perplexing problems arising when metal is heated to a melting point are fully explained and the proper methods to overcome them shown. 167 pages, fully illustrated. Price **$1.00**

CPSIA information can be obtained
at www.ICGtesting.com
Printed in the USA
BVHW082051201118
533619BV00011B/1673/P